Human Cloning Debate

4th edition

edited by

Glenn McGee Ph.D.
& Arthur Caplan, Ph.D.

BERKELEY HILLS BOOKS
Berkeley California

Published by Berkeley Hills Books
P.O. Box 6330, Albany California 94706

Distributed by Publishers Group West

Printed in Canada by Transcontinental Printing

ISBN 1-893163-65-2

Library of Congress Cataloging-in-Publication Data

Human cloning debate / edited by Glenn McGee and Arthur Caplan.— 4th ed.
p. ; cm.
Includes bibliographical references.
ISBN 1-893163-65-2 (pbk.)
1. Human cloning—Moral and ethical aspects.
[DNLM: 1. Cloning, Organism—ethics. 2. Bioethical Issues. 3. Religion and Science. 4. Stem Cell Transplantation—ethics. QH 442.2 H9178 2004] I. McGee, Glenn, 1967- II. Caplan, Arthur L.

QH442.2.H85 2004
176—dc22 2004013014

contents

PART III: PRO-CLONING VOICES

PART IV: GOD AND THE CLONE

Acknowledgements

We have in no small measure benefited from the generous commitment to reproductive bioethics research and public outreach at the University of Pennsylvania Department of Medical Ethics, including grants from The Greenwall Foundation, The Haas Foundation, and the U.S. Department of Education. The commitment of Rob Dobbin at Berkeley Hills Books to develop public debate about cloning has enabled this book to earn the distinction of being the best selling of its genre, and has made working on the 4th edition a pleasure. Thanks also to Roopali Malhotra, our colleague at the Penn bioethics center, as well as staff members Janice Clinkscales and Gloria Jones, for their assistance.

Introduction

Glenn McGee and Arthur Caplan

Late one February afternoon two years ago, one of us (Dr. Caplan) found himself sitting in front of a group of very attentive delegates at the United Nations headquarters in New York. They were there to begin a discussion of whether the UN should pass a convention to ban human cloning. Caplan had been asked to chair a special advisory panel which was to provide the delegates with both scientific and ethical perspectives on cloning.

In a way, this hearing was occasioned by the birth in 1997 of Dolly, the sheep made from the fusion of a black sheep's egg and DNA extracted from a mammary cell of a white sheep. The first cloned mammal was created in a sleepy Scottish village by a team of fairly even-tempered scientists, the leader of whom had adopted his own children and had no interest in replicating human beings. Dolly was named for a country music singer and displayed on the Today Show. Dolly was clearly born into a scientific class of her own, as you will read in these pages in essays by Potter Wickware and others. But the changing understanding of family and the changing ways we make families set the context for a debate about human cloning. Before Ian Wilmut and Keith Campbell at the Roslin Institute made their landmark discovery, couples and churches and mosques and synagogues and physicians and nurses and a broad swath of the public in Scotland, the U.S. and indeed the world over were already beginning to struggle with a larger debate about making babies in new and different ways.

The attention of the U.N., however, was directed only to cloning. The UN had made no progress toward enacting a ban. On the day Caplan spent listening to the delegates discuss cloning, it was clear that consensus would be very difficult to achieve. Many nations agreed that using cloning to make human beings is wrong and that cloning for reproduction should be prohibited. But, there is another reason to clone and that was hotly debated at the UN that day as well. Many of the delegates were not as familiar with the possibility of using cloning to make human embryos which might prove valuable in medical research as a source of human embryonic stem cells. This type of cell could possibly be used to create any number of cell types that people with injuries or diseases might benefit from receiving as transplants.

Some nations, notably China, India, Singapore, Israel and the United Kingdom indicated that inside their borders scientists were either already conducting research on cloned human embryos or would soon begin doing so. A few other nations expressed an interest in possibly hosting scientists who might conduct such research. Other delegations and observers, notably those from the United States and the Holy See, made it clear that any destruction of human embryos constituted killing and should be banned regardless of the benefits that such research might bring to the diseased or disabled. They argued for a ban on all types of human cloning.

In some ways the battle over cloning is difficult to understand. The science of cloning is in its infancy. No scientist or doctor of any international repute has indicated any interest in using cloning to make people. Pharmaceutical companies are not especially interested in trying to work to clone people since they see much bigger markets in the cloning of animals and cells. And no one really knows for sure whether stem cells derived from cloned human embryos really will prove useful as a way to cure diabetes, liver failure, Parkinsonism, spinal cord injuries or any other disease or ailment.

Human somatic cell nuclear transfer, otherwise known as the creation of a creature that bears great resemblance to an embryo through "cloning," involves:

- The starvation and subsequent implantation of DNA from specialized, non-sexual cells of one organism (e.g., cells specialized to make that organism's hair or milk) into an egg whose DNA nucleus has been removed.
- The resulting egg and nucleus are shocked or chemically treated so that the egg begins to behave as though fertilization has occurred, resulting in the beginning of embryonic development of a second organism containing the entire genetic code of the first organism.

On the other hand it is easy to see why there is so much interest and concern. There is seemingly no end to the parade of kooks, cranks, cultists and conmen who issue press releases proclaiming that they are close to success in cloning a human baby. Why? Proponents of cloning suggested it might serve as a new, unusual but perhaps efficacious treatment for infertility, enabling those unable to pass genes to future generations to do so in a way that is at least analogous to the familial linkage of twins. And, they point out, scientists have created animal clones and at least a small number of human cloned embryos with hardly any oversight or public accountability. There are, as we will read, obvious risks to any resulting offspring: mammalian cloning, through this nuclear transfer process, has resulted in the birth of hundreds of organisms to date. However, significantly more nuclear transfer generated embryos fail during pregnancy than would fail in sexual reproduction, and a substantial majority of cloned animals who have survived to birth have had some significant birth defect.

And for some who believe that any human embryo is a person from the moment of its creation, the fight over human cloning is a fight over both about what constitutes membership in the human community and about the morality of abortion. Many opponents of abortion hope that if they can gain legal recognition for cloned human embryos they can then move on to get legal standing for any human embryo or fetus.

One such person is George W. Bush. A few months after the hearing at the United Nations President Bush announced in a speech from the Rose Garden at the White House that he favored a ban on all forms of human cloning including the cloning of human embryos for the purpose of stem cell research (Bush 2002).

The President warned that in our zeal to find benefits and cures we could also "travel without an ethical compass into a world we could live to regret." Throughout the rest of his speech were salted terms such as "products," "design," "manufacturing," "engineered to custom specifications." The President was concerned that cloning would lead to the literal manufacture of human beings (Bush 2002). A few months later on July 10th the president's handpicked Council on Bioethics told the president what he wanted to hear—that moral concerns about human cloning were sufficient to warrant a complete ban on using cloning to make people and at least a four-year moratorium on using cloning for research purposes.

The President is hardly acting alone in sounding the tocsin of moral concern about the dangers of cloning. He is simply the most prominent among a long list of conservatives, pro-lifers, neo-conservatives and a small smattering of neo-green thinkers who see the cloning as holding the seeds of the degradation of humanity.

So is there a strong case against human cloning. Reproduction, or perhaps more accurately, replication of an organism's DNA identity does not normally occur in mammals, with the exception of twinning, which always results in the simultaneous birth of siblings. Only plants reproduce through replication from one generation to another. Would it be unethical for anyone to try to clone a human being today or at any point in the future? In addition to the obvious risks to the first child, debated herein, those who oppose human cloning point to the repugnance of a style of reproduction with such profound potential for vanity, arguing that the freedom of children and nature of the family are in danger.

And is it immoral to create cloned human embryos simply to destroy them for the purposes of obtaining stem cells to use in medical research? The case against human cloning may not be as strong as it initially appears.

There can be no debate about the fact that it would be irresponsible and morally wrong to try to use cloning to make a human being today or anytime soon. The experience using cloning to make sheep, cows, pigs and mice has made it abundantly clear that cloning is dangerous. There is real risk of death for the clone, a high risk of disability and for there are also very real risks for the surrogate mother who carries cloned fetuses to term. Without better safety data from animals including primates there is no ethical justification for trying to clone a human being.

But safety, while a very real concern, is not a concern about cloning per se. Presume that cloning were to someday prove safe. Would it still be ethically wrong to use it to make people? Any answer that pins the dangers for early prospective clones on something other than mere physical harms unique to the cloning process can capitalize on either of two conceptual problems:

- One is attempting to protect a future potential person against harms that might be inflicted by their very existence, and
- Societies around the world have indicated that they believe that the early cloning experiments will breach a natural barrier that is moral in character, taking humans into a realm of self-engineering that vastly exceeds any prior experiments with new reproductive technology.

Laws that would regulate the birth of a clone are philosophically difficult in part because they traverse complex jurisprudential ground: protecting an as-yet nonexistent life against reproductive dangers, in a western world that, in statutory and case law at least, seems to favor reproductive autonomy.

Many people seem inclined to put those philosophical issues, nonetheless, into a position of primacy in the human cloning debate, including the President and his chief bioethical advisor Leon Kass. But, the case against cloning when safety is taken out of the equation is a more difficult case to make than that which pivots upon safety alone. This is true whether one considers merely reproductive cloning or cloning for the sake of embryonic-based stem cell therapies.

One such argument against cloning people is that it is wrong to manufacture people. But cloning human beings is no less manufacturing them then using test tube baby technology or artificial insemination or even neonatal intensive care to make a baby is. No one feels any the less human for having been born in a neonatal unit or delivered by forceps or started up in a Petri dish. Clones would be no less people with free will and human dignity then any other person. Or would they?

Others worry that cloning is wrong because everyone is entitled to their own unique genetic endowment. This too is not a strong argument since identical twins and triplets already exist and do quite well despite having another person around with the exact same genes. Even if one is worried that parents will try to manipulate or force the clone to behave or develop in certain ways it has to be said that this is precisely what parents do with the children all the time whether they have a genetic tie to them or not. Should laws against cloning reach into social preferences about how children should be raised that are not enshrined in law?

Therapeutic Cloning

Even if the case against making human babies by means of cloning is not as strong as it might initially appear, perhaps the case against making human clones simply to destroy them for their cells holds? There are special problems with the arguments concerning a ban on this form of cloning.

Those who think it wrong to make and destroy human cloned embryos almost always believe that embryos are people from the moment of conception. But the fact remains that left in a dish in a lab a cloned human embryo has no potential for personhood unless one assumes the voluntarism of highly trained specialists and of women with empty wombs, and even then such embryos are only dubiously embryonic in that their potential to develop in a human uterus has been anything but established, and their differences from "ordinary" embryos—whether or not one considers such embryos to be persons—have been shown to be abundant and significant. So it is not self-evident that it is immoral to make and destroy cloned embryos on the grounds that this is the same as killing a human being.

The United Nations is still arguing about whether or not to ban cloning as are Congress and the President and many state legislatures. Ironically, using cloning to make people may turn out to simply be impossible since it may always be too dangerous and too risky for the baby. If that is so, a particular irony is that cloned human embryos might be the best source of stem cells for research since they might lack any potential for personhood. Either way, those who in this volume address questions of the moral status of the clone are part of what has become the kitchen sink of bioethical debates, involving as many obvious issues about cloning as one could conjecture, as well as a number of subtle arguments that depend on careful science and good public policy debate. How well society handles cloning will demonstrate not only how it handles its most extreme and extraordinary cases of conflict in medicine, but also how prepared it is for a world in which different kinds of personhood and parenthood may be as ubiquitous as new kinds of food and transportation.

A Roadmap

The fourth edition of *The Human Cloning Debate* is organized into six sections, covering the shape of the cloning debate, the relationship between personal and communal rights to control cloning, the emergence of revulsion about cloning and equally rapid emergence of bioethics to deal with fears and policy, the social and moral problems with choosing clones, the right to cloning treatments for the infertile, the rights of cloned offspring, and the role of religion.

In Section One, Potter Wickware deals with the science of cloning, highlighting the fact that reproductive cloning is widespread in nature, and tracing the history of mankind's experiments with cloning technnology starting over a century ago, and leading earlier this year to the production of the first human embryonic stem cell clones. In "The Ethics of Stem Cell Therapy" Glenn McGee et al. turn to face the challenges—clinical, legal, and philosophical—posed now by advances in stem cell cloning in particular. In "Human Cloning and Human Dignity," released in July 2002, The President's Council on Bioethics lays out the state of the cloning controversy in both its reproductive and therapeutic aspects (though the terms "cloning-to-produce children" and "cloning-for-biomedical-research" are favored instead), and the reader is carefully led through different cloning techniques with their respective ethical baggage. While reproductive cloning is roundly deprecated, on the question of "cloning-for-biomedical-research" a "spurious consensus" is shunned in favor of expressing the Council's collective ambivalence about the advisability of this research.

Part II culls some of the most considered and influential anti-cloning declarations to have appeared since Dolly first debuted, including several representative voices on the American right. Charles Krauthammer in "Crossing Lines" raises moral objections to any research that would culminate in the destruction of embryos, and advances a "slippery slope" argument to warn against extending propagation of blastocysts past the seven-day stage. Leon Eisenberg fashions a consequentialist argument, grounded in genetic science, to the effect that, whatever its promise among lower animals, reproductive cloning has a very dim future in human eugenics. In her philosophically informed essay "Reflections on Dolly: What Can Animal Cloning Tell Us about the Human Cloning Debate?" Autumn Fiester takes up the issue of what, precisely, cloning can contribute to animal husbandry, and argues that even here it appears to be a Pandora's Box. As to its applications to humans she concludes, "Given both the consequence-based and principle-based objections to animal cloning, it is evident that this science raises serious moral concerns and reservations that proponents of animal cloning must address." Sherwin Nuland argues that beyond institutions to stop cloning is the question of social maturity, which, in his view, Americans clearly lack. Leon Kass, responding to the "thin gruel" of bioethics, argues that our collective reticence about cloning technology ought to be granted more credence. His claim that our thoughtless embrace of technology might literally bear fruit in cloning figures prominently in many arguments in favor of a cloning ban.

Part III, "Pro-Cloning Voices," opens with a direct refutation by one of us (Arthur Caplan) of Leon Kass' reasons for opposing cloning science. John Robertson then makes a clear case for reproductive rights, arguing that decisions about some kinds of cloning technology might best be left to parents and individuals. Gregory Stock locates cloning and other emergent gene technologies squarely within the tradition of scientific advancement going back to the Enlightenment, and is optimistic, even fatalistic, about their eventual currency: "The question is not whether such technology will arrive, but when and where...and what it will look like." Ronald Bailey fashions a conservative pro-cloning position, arguing that it may be an appropriate technology if utilized in accord with the free market. Ian Wilmut and one of the editors (Glenn McGee) argue that the best model for understanding cloning and new reproductive technologies might be that of adoption, requiring couples to clear their intentions for cloning with a family court, and ensuring that society will support both the children of cloning and the families that it creates through this new science. Chris Mooney reports on the political status of the debate in Washington; he believes that even though both legislative houses are Republican-controlled, the trend is in favor of regulating, rather than banning, all types of cloning—a trend Mooney sees as a victory for common sense.

The arguments of a number of theologians and representatives for religious groups are presented in Section IV. While the authors certainly cannot claim to speak for all members of the religions they here purport to represent, each at least shares with their fellow contributors the view that cloning is an issue of deep religious significance requiring a new orientation to theological and religious texts.

Section V is devoted to the stem-cell offshoot of the cloning debate. Both pieces make the case that the discussion so far has been hobbled by a false assimilation of stem cell science to other issues or technologies—abortion, on the one hand, as we (McGee and Caplan) argue, and reproductive cloning on the other (so Vogelstein et al.). As the title of the latter's essay implies ("Please Don't Call It Cloning!"), they believe that the procedures in either case are so different as to indicate a different name than "cloning" for the technique of stem cell development.

Section VI returns to the focus of Mooney's piece: i.e. Washington DC, and how the issue of cloning in both respects is being addressed by the current administration. Included are two speeches President Bush has delivered articulating his position, the first on human cloning legislation, the second on stem cell research. Between them they have largely framed the current debate on cloning regulation, as Eric Cohen and William Kristol point out in the book's final piece, which goes on to speculate as to how that will play out with the electorate in 2004.

References

Bush, GW "President Bush's Remarks in Opposition to Human Cloning," Speech in the Rose Garden, Washington DC, April 12, 2002 (reprinted in Kristol and Cohen, 2002).

CNN Bush to allow limited stem cell funding, August 10, 2001: http://www.cnn.com/2001/ALLPOLITICS/08/09/stem.cell.bush/index.html.

McGee, G. *Beyond Genetics: A User's Guide to DNA* (Harper Perennial, 2004).

id., *The Perfect Baby, 2nd edition* (New York: Rowman & Littlefield, 2002).

Fukuyama, F. *Our Posthuman Future* (New York: Farrar Straus and Giroux, 2002).

Kristol W. and E. Cohen, eds. *The Future Is Now: America Confronts the New Genetics* (New York: Rowman & Littlefield, 2002).

History and Technique of Cloning

Potter Wickware

> Whereas ordinary mortals are content to imitate others, creative geniuses are condemned to plagiarize themselves.
>
> —Vladimir Nabokov, *Ada, or Ardor* (1979)

Nabokov's reflection strangely illuminates today's debate about human cloning. The debate commenced in February, 1997 with the historic announcement by Ian Wilmut, a Scottish cell biologist, that his group had succeeded in producing Dolly, a genetic clone of a six year old ewe, from breast tissue cells which had been preserved in a freezer. Immediately after Dolly was born there was a surge of speculation, both in cell biology and the world at large. Could the same be done with a human?

The question was answered in February, 2004, when W. S. Hwang and S. Y. Moon, of Seoul National University in South Korea, announced that they had created lines of clonal human embryonic stem cells, an accomplishment which unambiguously validates the premise of this book.

It's important immediately to distinguish between reproductive and therapeutic cloning. The object of therapeutic cloning, typified by the Hwang-Moon stem cells, is genetically identical cells. Stem cells are not intended to be—and in fact could not be—implanted and nurtured through gestation and carried to term. Rather, they live in petri dishes and serve in scientific experiments. They also have an exciting future as source material for replacement heart, brain, kidney and other types of tissue for patients who have suffered disease or trauma. Reproductive cloning's goal, by contrast, is genetically identical, full-grown animals for agricultural and pharmaceutical purposes.

Nevertheless, since human cells have been cloned, *could* they be implanted and carried to term, as animal cells have been? In theory, yes, although, as we'll see below, the high rate of birth and development problems in cloned animals provides a compelling biomedical reason—leaving aside the moral and ethical considerations that are explored elsewhere in this collection—not to undertake the attempt. Ian Wilmut, famous for cloning Dolly, the Dorset ewe in 1997, specifically warns against attempting to clone humans, saying that of the few embryos that might be created many would die early and the remainder would bear an unacceptably high risk of becoming abnormal children and adults.

But the mere possibility that a human being could be cloned introduces a sense of philosophical confusion in tandem with an inspiring scientific advance. The fact that cloning technology gives rise to both desirable outcomes and ones to be feared and avoided puts us in an ambiguous domain where features in the landscape have dualistic aspect of both good and bad.

Since the mid-twentieth century advances in bioscience have occurred with breathtaking—some might say frightening—speed. To the genetic engineers of the 1970s the idea of cloning an entire genome and raising a viable animal from a single adult body cell would have been unimaginable, as indeed it was to the vast majority of the scientific community prior to 1997. But today it appears likely that the intellectual velocity of the present will hold steady or even increase in the future, causing startling new choices to reveal themselves.

A good starting point for coming to terms with the choices, challenges and promises of cloning technology is to begin by understanding what cloning is, where it came from, and how it works. The rest of this chapter gives a broad, nontechnical overview of these points.

What is cloning?

First of all, just what is a clone? It's a simple word (Greek "klôn" means "twig") with a lot of particular meanings. An identical twin is a clone. So is a potato. So are the now numerous band of cloned animals. Cell lines like those of Hwang and Moon, and gene libraries used in medical research are clones. In nature reproduction by cloning occurs among many plants, in honeybees and wasps and amphibians. With the important exceptions of sperm and egg cells, all the cells in our bodies are clones, and the process of cloning is one that goes on continually. For example, skin cells and the cells that line the gut turn over rapidly, and genetically identical copies—clones—are produced to replace the ones that wear out. Similarly, when activated by a pathogen, the B memory cells of the immune system, which produce antibodies, quickly go into production and churn out identical clone armies to swamp the invader.

What these disparate entities have in common is that they are all genetic xerox copies based on a previously existing template or master. They are made without sex. They come into being without the reshuffling and reordering of hereditary material that takes place when two parents mingle their genes in their offspring, so that a child might end up with its father's nose and its mother's personality.

While it might seem strange to say that an identical twin originates without sex, in fact it is the sexless fission of the early embryonic clump of cells into two clumps that gives rise to a twin. That origin may be only one cell division away from the sexual event, but that single division, with its identical rather than reshuffled duplication of genes is, in reproductive terms, a yawning gulf of separation. For the potato, some time in the indeterminate past a sexual event occurred, which was followed by many rounds of clonal, or vegetative, duplication, so that a potato is about as sexless a creature as one can imagine. As for Dolly, the famous Scottish ewe, she was forty or so rounds of cell division and six years, plus some time in the freezer, away from the sexual event that gave rise to her genetic identity.

Cloning history

Animal cloning had slow beginnings about a hundred years ago in an esoteric and to outsiders unremarkable area of biology. In the nineteenth century Schleiden and Schwann, Sutton and Boveri, and Flemming, respectively, gave us the cell theory, revealed the existence of chromosomes, and described mitosis.

Following the trail blazed by these pioneers, Hans Spemann (1869–1941) devoted himself to exploring the influence of the nucleus on cellular development. In 1902 he did the world's first embryo-splitting experiment. Using a cinch made out of an infant's hair, he split a two-cell salamander embryo and found that each cell — clones of each other — subsequently developed into an adult salamander. This proved that early cells carry all the genetic information needed to generate a complete organism. He continued the work with later-stage embryos, and 1938 proposed the "fantastical experiment" of cloning by nuclear transfer.

In nuclear transfer the nucleus is removed from one cell and placed in a different one, typically an egg cell, whose own nucleus has been removed. Theory and technique advanced to the point where the operation was feasible and in 1952 R. W. Briggs and T. J. King at Indiana University carried out the experiment in frog cells. Success was evident in the viable tadpoles that resulted. Briggs himself was only dimly aware of the implications of his deed, declaring at the time, "Although the method of nuclear transplantation should be valuable principally to the study of nuclear differentiation, it may have other uses."

Cloning timeline (*NT: nuclear transfer. ES: embryonic stem cell*)

1838	cell theory, Schleiden & Schwann, Germany
1882	mitosis, Walther Flemming, Germany
1887	chromosomes, Theodor Boveri, Germany
1902	chromosomes linked to heredity, T. Boveri & W. Sutton, Germany & USA Walter Sutton
1902	slamander clones by embryo splitting, Hans Spemann, Freiburg, Germany
1938	Spemann proposes "fantastical experiment" later known as NT, Freiburg, Germany
1952	tadpoles cloned by NT from embryonic cells, R. Briggs & T. J. King, Indiana Univ.
1962	frogs cloned by NT from adult cells, John Gurdon, Oxford
1972	Paul Berg, recombinant DNA, Stanford, Califor nia
1973	first transgenic organism, E. coli carrying frog gene, by S. Cohen & H. Boyer UCSF, California
1984	first mammal cloned, a sheep, from early embryonic cells, by Steen Willadsen, Roslin Institute, Scotland
1986	first cow cloned, from differentiated embryonic

	cells, by Willadsen, Roslin Institute.
1995	sheep (Megan & Morag) clones from differentiated embryonic cells, Roslin Institute.
1996	sheep (Dolly) cloned by NT from adult cell, Ian Wilmut, Roslin Institute.
1997	transgenic sheep (Polly, carrying insert of human Factor IX gene), Roslin Institute.
1998	mice cloned by NT Yanagimachi & Wakayama, University of Hawaii
1998	human embryonic stem cells isolated by Thomson, Wisconsin
1999	transgenic goats (human antithrombin III in milk) cloned by NT from fetal cells, Tufts University
2000	rhesus monkey cloned by embryo splitting, Oregon
2000	pigs cloned by NT from adult cells, PPL Therapeutics, Virginia
2001	gaur (endangered ox) cloned by NT, Advanced Cell Technology, Massachusetts
2001	mouflon (endangered sheep) cloned by NT, Italy
2002	rabbits cloned by NT from adult cell, France National Ag Institute
2002	goats cloned by NT from adult cell, Nexia Biotech, Canada
2003	banteng (endangered cow) cloned by NT, Advanced Cell Technology, Massachusetts
2003	rat cloned by NT from adult cell, France National Ag Institute
2003	mule cloned by NT from fetal cells, University of Idaho
2004	Hwang–Moon human ES clones, National University, Seoul, Korea

Other investigators joined the growing throng exploring the cell and the technique continued to mature. Four years after Briggs and King, John Gurdon at Oxford University cloned frogs using nuclei from adult cells as starting material. The pace further quickened with the advent of molecular biology in the '50s and '60s, which provided new tools with which to manipulate the cell, as well as markers to track changes as they occurred. The DNA revolution also opened the door to transgenic clones, that is, clones with one or more genes from some other organism stitched in to its own chromosomes.

In the mid '80s cloned sheep were generated from embryonic cells, and a decade later Dolly was brought into being, using fully differentiated—that is, adult—material from a frozen tissue sample as starting material. In late 1997 Ian Wilmut's Roslin Institute one-upped Dolly with Polly, a transgenic sheep clone with a gene insertion for human Factor IX, a valuable blood-clotting protein, designed to be secreted in the animal's milk. More recently, cattle, swine, goats, cats, mice and other domestic animals, as well as several extinct or nearly extinct ones (banteng, gaur, African wildcat) have been cloned. The culminating event to date (at least from an *H. sapiens*-centric perspective) is the production of Hwang and Moon's human embryonic stem cell lines in early 2004.

To understand how cloning works it's necessary to grasp a handful of underlying concepts: DNA and heredity, proteins and chromosomes, and the cell cycle.

DNA, chromosomes

DNA is a polymer, chemically not unlike rayon or hair in that it is made up of fantastically long and prolific strands. DNA strands are punctuated by regularly occurring projections called bases, which stick out sideways from a long backbone, like pickets from the rail of a fence. Two strands fit together joined by their bases, so that the duplex resembles a railroad track, with crossties corresponding to paired bases. Human cells contain about 3 billion bases;

scaled up to railroad track size the DNA in each cell would be about 20 million miles long.

In one respect DNA is a boring molecule because it is so long and monotonous, but from another perspective it has a most remarkable property. During the brief period when cells divide, the double-stranded DNA comes apart down the middle, like an enormously long zipper. Then down each half-track comes DNA polymerase—a protein complex which challenges the writer for adequate superlatives—rather like a repair locomotive, and duplicates each missing half. When the process is complete there are two sets of DNA, one for the mother cell, and a new one for the daughter cell-to-be. The synthesis is a close to perfect one-to-one matching of base for base, and the DNA in the proliferating cells becomes the track that guides the individual through its lifetime, and its progeny down through the generations.

There are four kinds of DNA bases, A, C, T and G, named for the first letter of their respective chemical names. They occur in any order and make up a code that goes on interminably something like this: ...GAGATTTAACCGA... A small percentage of the stream contains stretches called genes, which at the right time and under the proper circumstances the cell is able to decode and translate into proteins.

There's a wonderful passage in Homer where demigod Proteus, in an effort to escape from Odysseus, turns himself first into a lion, then a boar, a serpent, a wave, a tree. Proteins' protean nature allows them to carry out functions of structure, control, signaling, transport, chemical breaking down and building up. Among the four categories of life molecules—the others being carbohydrates (i.e., sugars), lipids (i.e., fats) and nucleic acids (i.e., DNA, and its relative RNA)—proteins are the sleight of hand artists, the producers, the movers and shakers of the body.

DNA has to be organized. Imagine a film archive without reels, with the film strewn about in tangles underfoot, with no systematic way for playing it back. The chromosome is DNA's reel and shelving system. A chromosome is a DNA strand complexed with proteins which serve as spools. In all cells of the body except germ cells (sperm and egg), where they are single-copy, or "haploid," chromosomes come in pairs, i.e., they are "diploid." Humans have 23 pairs of chromosomes, sheep 26, mice 20.

Chromosome spooling is dynamic, so that the DNA is practically all put away when the cell is dividing—as though it were moving day at the film archive—while during G1 phase, the normal activity state of the cell (more on G1 and the cell cycle below), filamentous portions of DNA are partially unraveled from the chromosome and float out into the nucleoplasm. The so-called "expression loop" exposes genes relevant for activity in a given cell, while the ones which are not relevant are folded up and unavailable. A chromosome with an expression loop is like a tape cassette that has been "eaten" by the tape player, with a long rolled-out portion connected to a nearby wrapped-up one. During embryonic development there are lots of expression loops, with genes throughout the chromosomes being highly active, but soon after cell division commences the cells begin to lose their "totipotency"—the power to generate an entire organism from a single cell. The cells begin to specialize, to "differentiate," with whole regions of their DNA not relevant to their final function beginning permanently to shut down.

With complete differentiation in the adult, up to 90% of genes in any given cell type are apt to be permanently turned off. For example, genes coding for the proteins that produce the complex sugars collectively called mucus are highly active in the cells of the gut, but not at all in the cochlear cells of the ear.

Cell Cycle

Cells divide and make clonal copies of themselves: this is the essence of Schleiden and Schwann's cell theory. Looking at the process a little more closely we discover that it proceeds according to a stereotyped timetable. So the cell is like a clock, and every time the hands reach midnight there are two clocks, then four, and so on. The hours of the clock are called G1, S, G2 and M. G stands for "gap" (old nomenclature for the periods between mitosis and DNA synthesis), or alternatively "growth," S for "synthesis" and M for "mitosis."

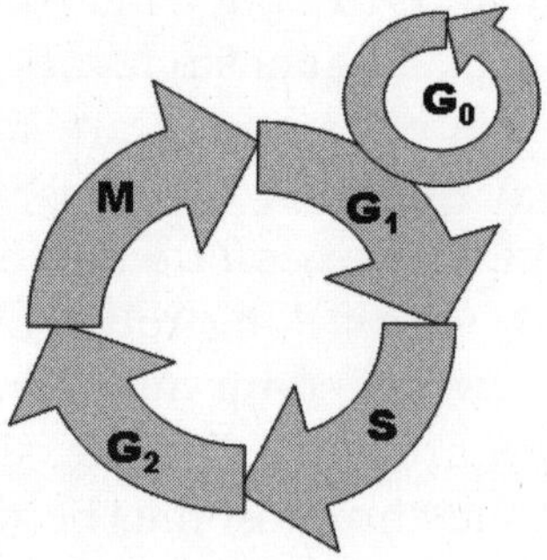

Fig. 1. Cell cycle. G1, normal gap/growth. S, DNA synthesis. G2 premitotic gap/growth. M, mitosis. G0, Same as G1 but without progression to S. (Potter Wickware)

In G1 the cell is open for business as usual. DNA germane to the cell's function is reeled out of the chromosome and accessible to RNA polymerases. RNA transcripts are fed to ribosomes, which churn out proteins. The proteasome-ubiquitin ligase system polices the cell for malfunctioning or worn-out proteins. Messages in the form of hormones and other signals from distant or nearby cells are processed. Myriad activities like these occur, and in general the cell hums with activity like a tiny city.

G1 ends when the cell gets a signal that a mitosis is in the offing. Then the everyday activities of the cell wind down and the specialized machinery of S phase — DNA synthesis — is turned on. When the huge copying job is complete and error-checked, a second set of chromosomes has been created and is ready to be passed along to the daughter cell.

S phase is followed by G2 phase, in which the cell makes preparations to divide. The mitotic spindles and centrioles form and the chromosomes condense. At length the great event of mitosis commences. All activities extraneous to the task are curtailed and the ballet of cytokinesis gets underway. At the end of the relatively brief, highly orchestrated process there are two cells where before there was one.

Some cells repeat this program for the life of the organism; skin cells and the cells that line the gut are examples. But other cell types - neurons and retina cells, for instance — permanently exit the cell division program after some developmental stage has been passed, and enter a resting state called G0. G0 is the same as G1 except that the cell never leaves this phase of the cycle, never synthesizes new DNA and goes on to proliferate. Never, that is, unless an error of regulation allows it to spin out of control, in which case a cancer or other pathology may result.

Nuclear Transfer

In a normal mating two sets of haploid (single-copy) chromosomes—one from mom, one from pop—fuse to become

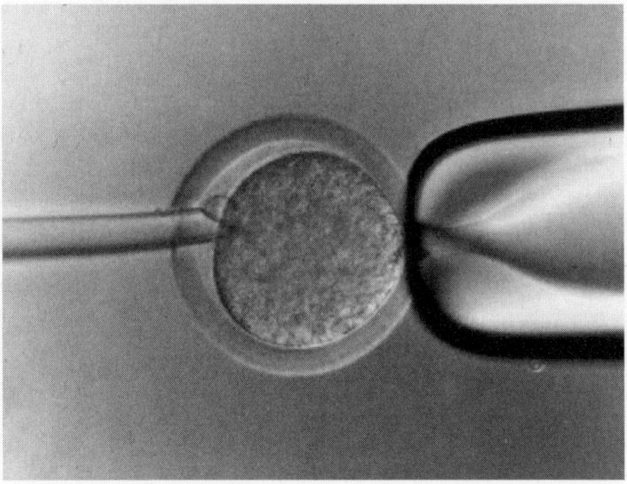

Fig. 2 Egg cell being microinjected (credit: PPL Therapeutics)

a single set of diploid (double-copy) chromosomes in the zygote. In nuclear transfer there's a diploid nucleus, but the fusion happened somewhere else and at some time in the past. Even so, if the placement of a diploid nucleus in the hollowed-out egg cell is done properly, the nuclear-transferred cell looks and acts like a zygote and development proceeds as if it were the result of a natural mating.

To look at the process in a little more detail, two cells are brought together, donor and recipient. The recipient is an unfertilized egg that's prepared by drawing out its DNA-containing nucleus, but leaving intact the outer membranes and the yolk—its accumulated store of nutrients and metabolic machinery. The donor is prepared by the inverse process: the nucleus is saved and the rest of the cell is discarded.

The donor nucleus is then placed in the outer region of the egg between the zona pellucida and the plasma membrane. (Think of a hard-boiled egg: the shell is the zona pellucida and the plasma membrane is the white film that lines the inside of the shell.) Tools used in the work include micropipettes, tiny glass tubes that are thinner than hairs. There are two types, a blunt-ended holding pipette to hold the egg cell by a mild suction, and the much thinner, sharply pointed insertion pipette for the drawing out and placing in of the nucleus.

Now that the nuclear membrane of the donor nucleus is in intimate proximity to the plasma membrane of the egg, the two elements are like soap bubbles with a common interface, but the nucleus is not yet inside the egg proper. Various methods—a pulse of electricity, a chemical shock—can be used to merge the two, mimicking the acrosomal process of fertilization that accomplishes the union in nature. From then on, if all goes well, the first cell divisions ensue.

Appropriate growth factors and nutrients in the serum prompt the cell to continue to grow and divide. It's left to develop in this way for about a week, to the blastocyst stage, by which time it has become a hollow ball of 64 or so cells, arranged in an inner and outer layer. At this point a crucial decision is made by the scientist in charge. If it is destined to become a full-grown animal it is implanted in a receptive female and, barring mishaps, carried to term. But if it is to become a line of cells, further manipulations are carried out, which are described below. Here we look at the fate of cloned embryos destined to become full-grown animals.

Nuclear transfer usually fails

Nuclear transfer leading to full-grown animals nearly always fails. Why? There are many observations, but at this point in a young science they're mostly empirical. When we get a good theoretical understanding of what's really going on in the cell when nuclear transfer takes place the difficulties seen today may diminish, but this understanding still lies in the future.

It is known that success, when it occurs, is due in part to correctly timing the cell cycle. Ian Wilmut succeeded after twenty-five years of effort where others had failed because he learned to synchronize the cell cycles of donor and recipient through serum starvation before bringing donor and recipient together. The cellular bread and water diet—a thin broth with minimal nutrients, vitamins and growth factors—slowed down the normally active donor DNA as well as protein expression in the recipient egg. In this way the two were able to properly mesh with each other and resume the program of the cell cycle as if an interruption had not occurred. In previous efforts, the overactive donor and recipient were out of synchrony with each other, and mistakes in cell divisions soon resulted in nonviable embryos.

Teruhiko Wakayama, at Japan's Riken Institute, who cloned mice in 1999, refined Wilmut's method by separating fusion and activation in time. First, with a micropipette the donor nucleus was placed inside the yolk sac of the recipient. Then, one to six hours later, activation was induced by putting the clone-to-be in a chemical bath that mimicked the acrosomal milieu of fertilization. This time delay, during which reprogramming (more on this below) takes place, combined with a scrupulous prevention of any mingling between donor and recipient cytoplasm during the transfer step, probably explains his lab's better than 2% success rate—better than Ian Wilmut's 0.4% success rate with Dolly, but still far below that of natural matings.

As Hwang and Moon announced in early 2004, they were able to achieve the first successful nuclear transfer in primate cells by tinkering with the nutrient and growth factor composition of the medium and incubation times. Their perhaps most significant innovation was to remove the egg nucleus not by drawing it out through a pipette, but rather by nicking the zona pellucida and letting it ooze out under its own pressure. Apparently, the gentler treatment minimizes disturbance of the microtubule network that holds the protein making and other metabolic machinery of the yolk in place—although many important details relating to the microstructure of the yolk are as yet unknown.

Reprogramming

Critical to a successful nuclear transfer, particularly if the donor nucleus is from an adult cell, is reprogramming its DNA. Most cells undergo specialization as they divide, becoming more committed to specialized function (liver, retina, skin, nerve, etc.), at the same time becoming less capable of universal potential. Until the advent of Dolly the accepted wisdom was that, in mammals, at least, it -as a one way process, but as is so often the case, the accepted wisdom turned out to be wrong. That it was possible to generate a complete animal from a single adult cell showed that cells could go backward in time after all, in the sense of being able to pick up functions that they had apparently lost. In the jargon of cell biology, they were capable of "retrodifferentiating," or of being "reprogrammed" to being capable of functioning like meiotic DNA again. In the little universe of the egg, donor DNA becomes infantilized, apparently losing memory of its former life. It happens during the nuclear transfer process, but scientists are tentative about many of the details of when and how.

At least two processes are involved: expression loops/chromatin and telomere length. As noted above, expression loops are active regions of the chromosome, brought about by chromatin proteins embedded in the chromosome that hold out some loops and hold others down. Deprogramming must erase the functional fingerprint that chromatin imparts to the cell.

Another substance likely to be involved with reprogramming is telomerase, a protein/RNA complex that acts on the telomere, a specialized region at the tip of the chromosome that serves both as "molecular bookend" and cellular clock. It is made from DNA patched together in a pattern of repeats, or "stutters," that define the boundary of the chromosome. Because of the way that DNA polymerase operates, a bit of the telomere is lost with every cell division. When the telomere is gone, after forty or so divisions, that's it for the cell. For a chromosome to get a new lease on life it needs its telomeres rebuilt. The agent responsible for rebuilding telomeres is telomerase; normally it operates only in germ cells.

Reprogramming mimics what happens naturally in germ cell development, with the important difference that processes that take months for sperm and years to decades for eggs occur within at most a few hours in nuclear transfers, between the time that nuclear transfer is completed and the onset of cleavage in the activated egg.

Safety of Nuclear Transfer

The low success rate of animal cloning by nuclear transfer is accompanied by birth and development problems among many—but by no means all—of the animals that do make it to term. Apparently the cloning process introduces subtle defects into the developmental programs of some of the clones. Damage to egg or nucleus may occur at various points: extraction of donor nucleus, injection into recipient, and the timing of the different steps as they are brought together. Incomplete reprogramming may also be to blame.

In 2002 a retrospective look at the health histories of cloned mice showed that most died prematurely, and seemed to be more prone than their normal counterparts to health problems such as liver damage, tumors and pneumonia, suggestive of some impairment of the immune system. One cloned mouse strain had a significantly shorter life span, tending to die young from liver disease. Another strain had a unique obese prepubertal phenotype. The offspring via normal breeding of that clone line were normal, however.

Similarly, a high rate of abortion and stillbirth in cloned cattle has been noted. One dataset reported by a US Food and Drug Administration report on safety of clones as food cited a cohort of 134 cloned cows in which 28 were stillborn, died, or were euthanized within 48 hours of birth. Hydrops, an abnormal water buildup in the fetus and the fetal membrane is quite common, and the animals tend to be larger than expected at birth for the breed. Newborn clones are also apt to be fragile in their first few days of life. Respiratory and cardio problems are also seen, as is a condition known as flexor tendon contraction, or bent legs. Qualitatively these phenotypes appear in all animals, but the numbers are higher among the clones.

Stem cells

Returning to the fork in the road where we left the week-old blastocyst, above, a radically different fate is in store if the cloner has decided it is to become a line of stem cells instead of an animal.

First the outer layer, which if the embryo were left intact would become the placenta, is removed. The cells of the inner layer, which would go on to become the fetus, are disaggregated and placed in culture, where they have a dual ability: they can reproduce indefinitely as is, or they can be steered down various developmental pathways to form all the cell types in the body. The Hwang–Moon cloned cell line is capable of forming bone, muscle and immature brain cells, for example. Hemopoeitic stem cells give rise to red and white cells and angioblasts, which in turn differentiate into the cells of the vasculature. Stem cells are even capable of switching lineages. For example, neural stem cells can give rise to hemopoeitic cells, and bone stem cells develop into liver or lung cells. In addition to embryonic stem cells adult stem cells are known to exist. They have a normal role in wound healing and in normal cell turnover, as in the gut, skin and hair.

Stem cells are important for culture-based studies of normal and abnormal human embryo development, for discovering new genes and testing drugs and for substances that cause birth defects. Perhaps most impressively, they are a renewable source of cells for tissue transplantation, cell replacement and gene therapies. Clinical targets include stroke, Parkinson's, Alzheimer's and other neurodegenerative diseases, diabetes, spinal cord injury, and repopulating the blood supply following radiation-based cancer treatments. Somewhat farther in the distance are universal donor lines genetically altered to prevent transplant rejection.

Why clone?

Despite its patina of modernity, and an association with a perhaps suspect scientific invention, cloning in nature probably predates sex as a reproductive mode. For plants, sex entails the production of flowers and seeds—as potentially risky and expensive a proposition for them as are the mating activities of humans. Specialized sexual organs, the flowers, must be produced, along with bright pigments, scents, and food sources to attract pollinators. Then follows the process of fumbling and groping by insect palps, bird beaks and bat tongues in their tenderest and most intimate regions. For all the enticing excitement of these events, consider the possible hazards: disease, injury, deformity, death. For all its plodding predictability, cloning is a more straightforward and economical way to accomplish the same result.

But cloning fails as a strategy during times of environmental challenge, for example when too-successful organisms outrun their range, or when some global catastrophe—an asteroid impact, an epoch of volcanism, some subtle shift in the chemistry of the atmosphere—strikes. Then creatures must adapt if they are to survive. That happens by sex. Sex produces offspring which are genetically different from the parent, with genetic resources which give them possibilities of response and behavior not available to the previous generation. According to the Darwinian principle, among the variable offspring will be some individuals better suited than others to changes brought about by an unpredictable environment. The more "fit" individuals, to use Darwin's word, have a better chance of surviving long enough to pass along their survival traits to their progeny. The process continues down through time, with species, families and orders waxing and waning according to the ease with which they are able to respond with new genetic combinations to changes in the environment.

A creature that lives in a steady state environment need not play the sexual game of chance and jeopardize a winning combination with needless recombination. This is not to say that plants, or anything else, can jettison sex altogether. Sex at least has to be held as a reserve strategy. Sooner or later just about everything that lives has to have sex. Even bacteria practice a form of it. But for much of the time, for many of the creatures that we share this globe with, sex need not be the common or preferred mode of reproduction.

Molecular cloning

Watson and Crick, who solved the structure of DNA in 1952, and their successors discovered that (with a few unimportant exceptions) DNA comprises the same alphabet and the same language in all organisms. From the discovery of the commonality of DNA it followed that it could be swapped from one organism to another, and that the source of the DNA—human white blood cells, for example—often made no difference whatever to the new host. Thus E. coli bacteria or yeast could be made to take up the human gene coding for insulin or Factor VIII, the clotting factor that is lacking in persons who have hemophilia, and treat it as if it were its own. Even more startling combinations were possible. In 1986, for example, a Japanese scientist, in an effort to produce a "reporter gene" for tracking inserts in gene transfer experiments, extracted the gene coding for the luciferase protein from firefly and inserted it in tobacco. When the plant was provided with luciferin, a chemical that emits light when acted on by luciferase, the tobacco plant glowed in the dark.

The by-now routine techniques used in molecular cloning is limited to comparatively short stretches of DNA making up the one or few genes that are of interest. The bacteria or yeast replicates its entire complement of DNA—its genome—for some number of times, and the stitched-in insert goes along for the ride. When the desired production is attained, the scientist somewhat ungratefully gives the yeast or bacteria a shot of lye to kill it, then snips out the now-abundant insert DNA, discards the rest, and proceeds to the next steps of research or drug development. Using this approach the genes involved in cystic fibrosis, muscular dystrophy, hemophilia, dwarfism and many other diseases have been discovered over the last twenty-five years.

Molecular cloning operations are rudimentary compared to the duplication of the entire set of tens of thousands of genes that must be carried out when cloning an entire genome. Molecular cloning might be thought of as running off copies of some small number of documents on a copy machine, while cloning by nuclear transfer to create a cell line or a full-grown animal is like duplicating the entire Library of Congress. But the two processes are the same in the essential respect that in both cases the molecule of heredity—DNA—is reproduced asexually.

Philosophical ambiguity

Returning to the position of philosophical ambiguity we

found ourselves in at the top of this essay, does not a feeling of uneasiness nag at us as we blithely, or even with great deliberation, make adjustments in familiar domestic creatures? Sex generates diversity. Cloning suppresses it. Natural systems are built on the basis of diversity, because the environment changes, and genetic recombination arose to deal with this unpredictability. The environment tomorrow may produce conditions that puts today's fittest creature at a disadvantage, favoring that organism which has a strange and possibly—under other circumstances—lethal gene-expression pattern. There's no way to see into the future, says the logic of nature, and no way to prepare in advance for its onslaughts, and so nature creates vast numbers of possible combinations, profligately, indiscriminately, with no respect for any particular individual. That nature makes no assumptions about the future may be good or bad, or irrelevant for the individual. It may or may not be a good thing that the naturally fertilized creature is unique. It may be unique by virtue of an unpleasant temperament, or disease susceptibility. Nature doesn't know, doesn't care. All that's certain is that circumstances will change, and what's a liability in the here and now could be a lifesaver in the there and then.

Certain human genetic diseases illustrate this point. Sickle cell disease, a form of anemia in persons of African descent, confers protection against malaria, and so it is a health advantage to carry a copy of this "disease" gene in malaria country. Similarly, cystic fibrosis, a choking disease of the lung that is relatively common among Caucasians, is thought to have conferred protection against the severe diarrhea that afflicted the people who lived in the filthy settlements of pre-modern Europe. Other genetically determined conditions such as color blindness or albinism have no known benefit, and afflicted persons' lives are to some extent impaired. Too bad for them, but conceivably the environment could change in some way in which these traits could be advantageous. Were that to occur, the individuals bearing them would survive and pass them along to their offspring, while those with what we deem normal skin pigment and vision would die out.

Cloning completely upsets this logic. Indeed we can see into the future, says the cloner, and as far as the clone and its environment are concerned we predict it will look exactly as it does now. Furthermore, the cloner has regard for the individual and rejects the group, with all the broad combination of possibilities created by random gene shuffling. Cloning is a complete reversal of the natural situation. It's as though you as the clone are playing a card game in which the same hand is dealt again. But in reality it's not the same hand, because all the other hands at the table are different, and this may well be to your disadvantage as a clone. Before when you were playing you won the pot with a full house, but this time around the fellow sitting across from you may have four of a kind.

This brings us to the final point, that a clone, despite its genetic identity, will not be identical as an individual to its progenitor unless an impossible condition is satisfied: it must be raised in an identical environment. Although its genotype is identical, its phenotype (Greek, "what is shown") may be spectacularly different. In bees, future queens are produced by providing a special diet high in sugar and vitamins to a worker egg or larva. In combination with chemical signals called pheromones, the special diet activates sets of genes that cause the insect to develop appearance and behavior that are spectacularly different from the brood mates she left behind. The queen has a barbless sting, is larger, lacks wax and other glands as well as the pollen-collecting apparatus, and is a prolific layer of eggs, in contrast to the workers, which lay no eggs.

Humans early in their development—clones or not—are affected and molded by their individual environments, and small stimuli result in large outcomes. From the very beginning the chemical environment in one uterus varies from that of another, and throughout childhood small emotional and physical influences with large long-term ramifications continue to accrue. Thus even though an organism may be genetically identical, it is not identical as an individual, because of the dynamic tension exerted by the environment. But even though through this reasoning we persuade ourselves that a cloned Jack the Ripper might turn out to be a great philanthropist, and a cloned Einstein a hopeless dunce, it's inescapable that these clones would have a smaller range of possibilities open to them than humans who attained their lives in the conventional way. These are some of the contradictions and complexities we face as we decide if we should welcome, tolerate or forbid the further development of cloning technology.

The Ethics of Stem Cell Therapy

Glenn McGee, Pasquale Patrizio, Vanessa Kuhn, Claire Robertson-Kraft

It is, perhaps, the most important scientific advance in the past 100 years and its potential is not even close to realization. It is the most controversial technology imaginable, an improbable combination of the abortion, cloning, fetal tissue, transplantation, gene therapy, animal rights, and enhancement technology debates, raising worries about women in research, sex, the regulation of in vitro fertilization (IVF) clinics, the danger of changing the human germ line, and the war against aging. Before it is developed, some of the most powerful politicians on earth will find themselves forced to modify deeply entrenched views, and a few dozen scientists will become billionaires through patents on bits and parts of embryos. More than 150 million Americans and perhaps another billion people around the world may be treated with it before the decade ends, yet almost no significant research involving human subjects has been performed with it. It commands the attention of the major newspapers, news media, and scientific and business press every day, yet not a single book has been written about it. It is the human embryonic stem (hES) cell, perhaps the most important innovation in the history of humanity's quest to understand its own origins, and key to dozens, perhaps hundreds of advances in medicine. It is also the most controversial technology imaginable, viewed by many as a Faustian bargain involving the trade of innocent potential lives to extend other lives, and by many more as a necessary sacrifice of part of human procreative mystery in the interest of curing disease.

Scientific Milestones on the Road to the Stem Cell Therapeutics Debate

The work that led to the development of a conceptual identification of what are today referred to as stem cells and to the development of research programs around the world, has early origins. In 1963, Bob Edwards, Robin Cole, and John Paul identified stem cells from cleaving embryos and blastocyst inner cell masses.

Cell lines developed from cells isolated from inner cell mass would not stop dividing and retaining their genetic characteristics, even after cryopreservation for several years. If blastocysts were cultured intact, Edwards and colleagues found that their inner cell mass formed colonies on trophectoderm and produced blood islands, nerves, muscle, connective tissue, phagocytes, etc. Richard Gardner, a student of Edwards', produced the first chimera following the transfer of a single stem cell into a recipient mouse blastocyst. It was then seen that stem cells colonized all tissues except trophectoderm. However, to gain full advantage of the potential therapeutic developments from the isolation of the stem cells, it was first necessary that significant research using human cells be undertaken. This was among the goals of Edwards, the pioneer of human IVF. The work of Edwards in bringing to fertilization and implantation the first human embryo through IVF, later to be born as Louise Brown in 1978, was most significant in this regard. In clinical settings, the birth of the first "test tube baby" was precursor not only for the treatments of infertility, but also for preimplantation genetic diagnosis (PGD), and chromosome studies. More importantly, in laboratory settings, it provided the opportunity for embryo research as grounding for what is today stem cell research.

At the same time another of Edwards' students, Peter Hollands, used stem cells from mice to colonize lethally X-irradiated mouse recipients and showed how they migrated through liver to bone marrow, spleen, and perhaps elsewhere. They colonized and became active within 3 to 6 days in recipients, saving mice from an earlier death and sustaining them through a complete life span. No cancers, inflammation, or damage occurred in recipients, and rat stem cells were as effective as mouse stem cells. In the early 1980's, Edwards and his group attempted to move this work forward by trying to get human stem cells from few spare blastocysts, but these experiments were stopped due to ethical concerns, rather than a failure to progress.

In 1998, Drs. John Gearhart and James Thomson published the identification of the pluripotent hES. Long before any clinical demonstration that hES cells could have therapeutic efficacy in the treatment of human disease, many scientists, advocates for those with degenerative disease, and politicians spoke and wrote of "the profound potential" of stem cells for medicine. Those who object to abortion, fetal tissue research, or IVF on moral grounds have condemned embryonic stem cell research and treatment in the strongest possible terms, advocating instead the use of stem cells derived either from adults or from blood obtained from the umbilical cord. The scientific facts that would make clear whether adult-derived or embryo-derived stern cell therapy would be most efficacious are not yet in evidence, yet both pro- and anti- hES arguments in the clinical and bioethics literature have focused on science. There is thus much confusion about how the scientific facts of the matter relate to underlying moral concerns. Both sides have sought middle ground, albeit largely without success.

From the point of view of consumers, activists, and patients, stem cell research may seem to have materialized from nowhere, a miraculous discovery with great potential. Unlike contemporary genomics, which has become decidedly goal directed and focused in character, the laboratories of stem cell research had not one or two therapeutic goals but in fact hundreds of possible research and clinical trajectories for their laboratories. Moreover, stem cells have long figured prominently in basic research in human and veterinary cell biology, in clinical trials of possible therapeutic techniques, and even in a number of successful therapies. Basic research involving stem cells is most often focused on fundamental problems of developmental biology, for example, how it is that specialized cells come into being, and how groups of specializing cells come to participate in coordinated activities. Basic stem cell research thus focuses on the time in, manner through, and extent to which somatic cells specialize during the development of an organism, and the role of stem cells in repopulation and repair of damaged or otherwise depleted cells in the mature organism.

Embryo Research as a Grounding for Stem Cell Research

For the purpose of this chapter, an embryo is the developing organism, understood to exist from the time of fertilization until the fetal stage as described earlier. Human embryos became broadly available for research purposes only following the development of IVF, developed in the 1970s by Steptoe and Edwards primarily to treat infertility. In 1978, Steptoe and Edwards documented the first birth through IVP; 4 years later, they reported their intention to freeze spare embryos for possible clinical or laboratory use. Since that time, scientists and clinicians have made use of embryos for solely research directed purposes. At one level, it has been noted that embryos are the centerpieces of anatomy and pathology research concerning the basic units of a process of development. This research is also demonstrably useful in improvement of the clinical efficacy of IVF, and for the investigation, at another level, of the diagnosis and treatment of hereditary and other diseases and injuries with the aid of preimplantation genetic diagnosis (PGD).

The roots of stem cell research are to be found in understanding the chain of events and set of structures involved in processes of embryonic and fetal development. At the root of this interest is the question of how a human embryo transforms into a complex human being. There are at least two kinds of hES Cells; best classified are totipotent cells and pluripotent cells. The totipotent hES cells are found in the dividing fertilized egg. These cells have the unique ability to develop into any cell or tissue types found in the human body, for example liver, cardiac, nerve, or blood cells; in addition, they have the capacity to form a complete organism. Pluripotent hES cells are found in the inner cell mass of the blastocyst: at the stage of development in which the dividing cell mass forms the shape of an almost hollow ball. While pluripotent human hES cells can develop into many if not all cell and tissue types, it is not currently believed that they would have the ability, if implanted in the human uterus, to divide and mature into an organism. Pluripotent stem cells are the cells most often used in embryonic stem cell research. In order to obtain embryonic stem cells, the inner cell mass of a blastocyst must be isolated from its outer shell, removing the embryo from what would have developed into the placenta. Moreover the inner cell mass is disassembled by taking out individual embryonic stem cells for research purposes. The embryos used for hES cell research usually come from embryos created through IVF but not used for that purpose. The euphemism "spare" or "leftover" embryo has been coined by clinicians and used by politicians to describe this source of cells for hES research and therapy.

New Concepts of the Clinical Utility of Embryonic and Stm Cell Research

Although it was not completely clear at the time of Edward, Gearhart and Thomson's publications exactly what would result from the identification and cultivation of pluripotent hES cells, it was immediately apparent that their findings had great importance both for basic and clinical research in humans and animals. First, a key discovery was the identification of a crucial point in the development of the human embryo at which the DNA in the nucleus of particular, undifferentiated cells no longer has the power to make another identical organism, the point at which totipotency is definitely not present. Second, and more important, these cells' nuclei can produce a wide range, and perhaps all, of the kinds of cells that populate a developing or mature human organism. Third, it is possible to derive these cells from the embryo, and to isolate them from other cells. Fourth, once derived, these isolated pluripotent hES cells can be cultured and frozen, transported and grown, fed and measured in a variety of ways. Fifth, these cells can be induced to produce differentiated cells. These cells might then themselves produce more cells that might be transferred from culture into the bodies of patients to replace a wide variety of damaged cells, or to perform a range of other tasks, from inoculation to the destruction of cancerous tissues to the delivery of drugs.

Several well-publicized clinical trials involving the transplantation of fetal tissue into patients with degenerative diseases of the brain and nervous system, such as Parkinson disease, have been conducted without success, despite successes using almost the same modality in mouse and primate trials. Although these trials did not specifically measure the activity of stem cells, they raised basic questions about the utility and toxicity of immature cells for transplantation. Clinical research that specifically involves stem cells has included a wide variety of tests of the effectiveness of transplanted stem cells in repopulating certain needed cell types in patients with, for example, bone cancer and diseases of the immune system. Techniques already in use include the harvesting of stem cells from umbilical cord blood and the transplantation of stem cells for the treatment of leukemia.

Enthusiasm about embryonic stem cell research quickly led to a larger discussion of the future of the work, and the implications of stem cells for broader debates about how to allocate health-care resources, how to proceed with caution in new areas of clinical research, and how to regulate research involving embryos, fetuses, or abortion. Wide calls for governmental investment in stem cell research were entertained both as part of the 2000 presidential campaign in the United States and as part of governmental hearings the world over. It was noted in the United States and elsewhere that, like mammalian cloning research, researchers whose work was funded by small companies rather than national or regional governments were making the most innovation in stem cell research. Arguments for government funding of stem cell research were usually linked to the claim that government funding would enable regulation, and if necessary restriction, on stem cell research. This argument received the endorsement of many ethics advisory boards, including, for example, the U.S. National Bioethics Advisory Board (NBAC), an arguably partisan board of ethicists appointed by President Clinton. What did not emerge immediately was the question of how patents filed in association with Gearhart, Thomson, and others might make it difficult for the government to exercise as much regulatory authority on groups like the NBAC, the American Association for the Advancement of Science (AAAS), and others that sought research leadership.

Ethical Issues in Contemporary Stem Cell Research and Therapeutics

Although the subject of research on embryos presents a variety of ethical and legal issues, the central issue in the Western debate has long been the moral status of the embryo. The debate over the moral status of the embryo is not unique to 20th and 21st century scholarship in science and bioethics. On the contrary, this controversy originates and is deeply rooted in different religious and philosophic views. In the Western philosophic tradition, the debate over the status of the embryo can be traced to Aristotle, who wrote of the ensoulment of the human at a particular stage, as did the pre-Socratic philosopher Heraclitus before him. Religious views of conception have been extensively debated in Judeo-Christian and Muslim scholarship dating to the earliest religious texts in those traditions. The contemporary question of the moral status of the embryo emerged during the U.S. controversy over the legality of abortion in the 1960s through 1980s, and continues to be an issue in the discussion of the use of most reproductive technologies today. Views on the moral status of the human embryo normally take one of the following three forms:

1. The human embryo has no intrinsic moral status; it derives its value from others.
2. The human embryo has intrinsic moral status, independent of how others value it.
3. Embryos begin with little or, no moral status and continue to achieve more and more status as they develop.

The position that an embryo has no moral status can be argued in several different manners. Because the fetus fully depends on the pregnant woman for development, many ethicists believe that it cannot be viewed as a unique entity. Many others—most famously Massachusetts Institute of Technology philosopher Judith Jarvis Thomson—argue that the best metaphor to describe the status of the fetus is that of a parasite (whether desirable or not), possessing no moral status independent of the mother. Those who hold this position do not object to embryo or fetal research on the grounds of the moral status of the fetus, and would refer to, for example, fetal surgery (whether conducted *ex utero* or *in utero*) as a procedure on the mother. The concerns expressed by those who hold this position about embryo research are focused on the long-term social implications of embryo research for the status of born persons, particularly those with disabilities. However, it is not held that the destruction of an embryo is inherently morally problematic.

The position that the fetus has intrinsic moral status is grounded in the view that a person is created at a moment in time that can be linked both to the consummation of an act by those who participate in its creation, and to the physical and legal initiation of that person's participation in the human community. The metaphor most often used to describe the status of the fetus for these purposes is that of baby. The ever-increasing presence of the fetus in public and private life has contributed to the view that from the moment of conception a person can be identified, independent of the risks that face a person so defined, and regardless of the plain differences between such a person (e.g., in the case of a frozen embryo) and a person who participates as a baby, child, or adult in the institutional life of the community. Given this view of conception and the embryo, the use of an embryo for research purposes is exactly tantamount to the use of any other vulnerable subject in research without consent, research that poses not only a great risk but in many cases has the dearly anticipatable outcome of death for the subject.

A variety of philosophers and scientists have argued for a developmental model of the moral and legal status of the human embryo and fetus, beginning with the claim that clinical changes in the embryo and fetus have moral significance because they represent, if not ensoulment, the development of concomitant ability of the being to participate in the human community. One way in which this position has been expressed is in the Roe decision, which held that pregnancy could be divided into three periods, corresponding to the degree to which the embryo has developed. The Supreme Court ruled that these periods represent the increasing standing of the emerging human person in the human community. Contemporary neonatal technology has made it possible to construct a clinical definition of viability, a time at which the developing fetus would be able to survive outside the womb. Important, though, is the fact that not only does the fetus change over the course of pregnancy, the technologies of neonatal care evolve as well, so that in the course of 5 years the moral status of a 22-week fetus would change with the state of the technology, rather than remain fixed at some natural point in development. Those who hold that the development and viability of an embryo is morally relevant to research all embryos, fetuses, and stem cells must face an interesting array of problems: how can values (e.g., the rights of the embryo) be derived from facts? What moral status is conveyed to a laboratory creation, for example, an embryo-like creature made from parts taken from several different species? What if any moral standing does the specialized cell in an adult have if it can be demonstrated that all that is required to turn that adult cell into a cloned embryo is a jolt of electricity or bath of enzyme? This position, held by the majority of registered voters in the United States and United Kingdom (as demonstrated by polls), is in many ways the most complex in virtue of its attempt to be responsive to changing science.

Law and the Ethics of Embryo Research

The moral issue surrounding embryo research leave the status of the embryo highly contested. The difficulty for the law is that when dealing with this terrain, it is fraught with confusion and presented to the courts in the form of a particular case, and often in a context where little expertise is available or admissible on the subjects of the science and ethics of the matters at hand. Moreover, in: courts, as opposed to the institutions of religion and philosophy, some consensus must be reached in order for the institution to complete its appointed task in each case. Even when the Supreme Court of the United States passes over a case on abortion, it has taken an, action that is both important for the purposes of allowing others to determine where the law stands, and for the purpose of the completion of the judicial task. This is inherently problematic for a number of reasons: not only does relatively little agreement exist between scientists, ethicists, lawyers, lawmakers, and religious leaders in regard to the status of the embryo, but it is also unclear what path embryo research should take from the perspective of the "experts about the matter," because in a legislative, judicial, and economic leadership vacuum, it is difficult to determine who the experts are.

The lack of consensus about the status of the embryo and the morality of research has resulted in what might be contradictory and unclear legal definitions in the United States at the state and federal level. Because it is extremely difficult to define the status of the embryo and the question remains hotly contested, most of the legislation tries to steer away from making a definitive statement that would outrage either side of the debate. The legality of embryo research varies too from country to country.

Experimentation on the embryo for the purposes of developing stem cell and other technologies, and for general knowledge, is legal in the United Kingdom and three Australian states under certain circumstances. In Germany, embryo research is banned completely. In the United States, debates over the legality of embryo research tend to pivot on prior state court holdings, federal agency rules and directives, or state laws on the status of the embryo. Even though the courts already attempt to resolve the debate over the status of the embryo, it must also undertake a new set of questions: should experimentation be allowed at all? If so, under what circumstances should it be prohibited? For the majority of U.S. residents, at least, polls show that some experimentation is desirable, so the question the courts face in the political arena is how the line between acceptable and unacceptable experimentation should be drawn.

Law and the Status of the Embryo

Historically and under common law, the fetus has not been legally protected until after complete separation from the mother's body. This view holds that because the fetus is not independent *in utero*, it cannot possess individual rights. Consequently, any harm caused to the fetus *in utero* has not been legally protected. Recent decisions criminalizing the termination of pregnancy or even activities that might result in eventual harm to a potential future person under certain circumstances have altered the tradition of common law concerning the fetus and embryo, as have lawsuits concerning wrongful birth. Specifically, mother and child are now able to make a tort claim for malpractice that takes the form of medical negligence if predictable harm to the embryo *in utero* has had a negative effect on the newborn child.

For purposes of defining the status of the embryo, courts also have relied on the personhood test: when, and under what circumstances, and given what kind of creature is an embryo considered an embryo, and when is it considered a person for the purpose of legal protection? In Roe v. Wade

(1973), the United States Supreme Court denied that the unborn be considered "persons" under the 14th Amendment. However, they failed to set forth a clear definition of personhood or explain why they denied the unborn such status. Consequently, there has still been much debate concerning the legal status of embryos, even given the aforementioned court interest in recognizing three periods of pregnancy and the clinical and legal significance of the third period or trimester for assigning increasing interests (if not rights) to the fetus.

Because Roe v. Wade did not clearly define personhood, the Court had to use other means to construct a definition of an embryo. This task, as most involving embryo experimentation, was and remains highly problematic. It is a task that has been taken up in many nations and states, one contingent on whether fertilization should be assumed to confer individuality and, if so, if fertilization is an event or a process. It is due to this debate that the courts, their advisory bodies, and legislation, have come to focus on the metaphysical question of identity, and whether or not personhood or individual identity ought to playa part in determining at what point an embryo is too mature (and thus possessed of moral standing) to be subjected to involuntary (thus, any) testing.

The Warnock Committee published a report in 1984 stating that destructive embryo research should only be permitted up to 14 days into development. The 14-day limit was based on the following argument:

Twinning can occur up until 14 days of development.

If twinning is still possible, then an embryo cannot be considered an individual.

Only individuals can have moral status.

Beings without moral status have no right to be free from destruction and thus can be experimented on.

The 14-day rule rests on the assumption that being an individual confers moral status on a being and provides its own definition as to when this individuation occurs. But other standards also have been proposed. One is the constantly evolving notion of viability: perhaps the viable fetus has moral standing, white the fetus that cannot survive outside the womb does not. Another is the standard of birth or even of informed consent with parental surrogacy, which would either rule in or rule put embryo and fetal research depending on one's (or one court's) view of the importance of informed consent or the nature of surrogacy for a fetus. This question has been raised in fetal surgery. Still another is the assertion that at conception or fertilization there is a person in place, but here the question remains at what moment the actual fertilization or conception takes place, and under what circumstances one could perform any clinical, or research procedures on a conceptus, and by whose authority.

Foundations of Ethical Debate in Stem Cell Research and Therapy

It has been maintained that not only does one's position on the ethics of stem cell research depend on the question of when conception occurs and what bearing each developmental milestone has on the moral standing of a fetus, but also on the underlying view one holds about values and ethics. One's ethics will determine the horizon of the moral inquiry. One's view about whether a moral matter tends to involve personal choices by involved actors who are rational, or is instead a broader and more social dialog leading to either a social contract or the creation of social institutions, will bear on whether one is willing to or capable of engaging in deliberative democratic discourse on this complex set of questions. There are several, theoretical questions in ethics that are of this variety.

Theory of Rights. It is claimed by some that because the embryo has no "interest" in living, it does not posses any right to live. This argument rests on the assumption that killing is wrong because it deprives a person with an interest in life of his or her necessary interest in life. If an embryo is neither conscious of life nor cares for the duration of its own, it has no intrinsic moral status under the theory of rights articulated by Robert Nozick and others. Specifically, it has neither a positive right to be thawed out from a nitrogen tank and given a womb, nor a negative right against being demolished while proceeding through development ensconced in the womb. The emphasis is on liberty interests, attached to the idea that a person is rational and capable of articulating interests, an emphasis with a number of weaknesses and strengths when elevated to a legal and moral argument.

Consequentialist Theory. For the consequentialist, the ends determine an action's moral status, and good ends justify the means necessary to achieve those ends. Embryos can be experimented on or even destroyed, consequentialists have argued, because the ends of embryo research outweigh whatever damage is done to embryos-including the destruction of embryos. This is justifiable as long as it is clear that the embryo's suffering or death is not less morally desirable (to itself or to others, understood in a variety of ways) than is the suffering of the patient, community, or family affected by a treatable or potentially treatable disease under investigation that uses stem cells that require the destruction of embryos.

Religious Views. A number of religions express views about abortion and indeed about reproduction and research, and these matters have been debated in intra-denominational and social forums. It is important to take note of one view that was held by the Vatican since 1859, because that view is in play in the political debate more than any other in the West. It is the view that the embryo obtains moral status at the moment of fertilization. Recently the Vatican has gone so tar as to link fertilization and moral standing to genes: with a unique genetic makeup, an embryo is given a soul. Because twinning can occur up until the 14th day of development and two zygotes can fuse, a theory of individual ensoulment predicated on genes and fertilization faces scientific hurdles no less than other views.

The Derivation Dilemma

Whatever its religious or scientific underpinnings, the ethical debate surrounding hES cells has recently centered on how the hES cells are derived and on whether or not they should be protected from destruction, much like an adult is. Using leftover IVF embryos for the purposes of hES cell research raises complex questions about the status of the embryo, the value of human life, and whether there should be set limits regarding the interventions into human cells and tissues. Furthermore, questions about adequate informed consent, oversight, and regulation also come prominently into play.

Those who support hES cell research argue that an embryonic stem cell, even though it is derived from an embryo, is not itself an embryo and thereby would never continue to develop into a fetus, child, and adult. Each stem cell is only a cell that can be triggered to become a specific kind of tissue yet could not be triggered to become an individual. Furthermore, the embryo at the blastocyst stage has not developed any kind of nervous tissue; thus, extracting individual stem cells would not be painful for the embryo. Since the embryos used for stem cell research come mostly from the leftover IVF embryos, which would otherwise be discarded, the proponents of stem cell research argue that it is better to use such embryos to find cures for debilitating diseases than to discard them, benefiting no one.

It is also argued that many of the embryos used to make hES cells are not embryos at all but instead something else, either "pre-embryos" or merely partially human cells. In many cases, no conception occurs in the creation of these cells, for example, in the case of nuclear transfer to make a genetically identical embryo-like human that grows to blastocyst but might not be able to survive implantation in a womb. What is an embryo, and what does it mean to make something that behaves like an embryo but could not come to term in a womb?

One attempt to resolve the debate over stem cell research involved the suggestion that researchers might obtain stem cells from embryos without actually engaging in the destruction of those embryos, or, perhaps more correctly, that the activity of embryonic cell research could or should be viewed as morally distinct from that of obtaining cells through the destruction of embryos. It also was suggested that totipotent cells might be removed from four- or eight-cell preimplantation embryos destined for IVF (without destroying the embryo, a technology performed with some frequency in contemporary reproductive therapeutic settings for purposes of preimplantation genetic diagnosis). This was originally proposed by the U.S. National Institutes of Health under the Clinton administration, and was in substance taken up by President George W. Bush. He suggested that although it is immoral to destroy embryos, some hES cells have already been derived from previously destroyed embryos, and the matter of the availability of those cells can be considered distinct from the matter of creating new cells through the destruction of additional embryos. He thus decreed that only stem cells derived from embryos destroyed before his speech would be made available for federal funding. As the President framed his compromise, "only those cells for which the life or death decision has already been made" would be eligible for use. He noted that 66 stem cell lines have already been obtained from embryos, "more than enough" to allow that research to proceed.

Predictably, a number of concerns were raised about the

President's rationale and about his policy. However, the overriding question was whether enough embryonic stem cells in fact exist. The issue of the suitability and scarcity of hES cell lines already derived at the time of Mr. Bush's speech called attention to the fact that many hES cell lines are subject to U.S. and international patents, and that many of the innovations necessary to derive, culture, differentiate, or otherwise manipulate stem cells are also subject to patents. However, should stem cells, embryos, embryo-like organisms, or the cells derived from them be eligible for consideration as intellectual property, whether through patents or other protections of law? Did President Bush compromise the principle that life begins at conception, making a political attempt at consensus, or did he merely address the political reality of overwhelming support for the research set against an incredibly vocal minority opposition that constitutes the bulk of the conservative party?

Another central problem is the permissibility of making embryos specifically for research purposes. There are two different types of embryos used: those classified as spare embryos that are left over from unsuccessful IVF and those cultivated specifically for purposes of being tested. Some people have ethical concerns about both of these methods; however, those who support research are more likely to question the ethical nature of the second of these two alternatives. The argument that it is acceptable to use spare embryos but not to create embryos specifically for that purpose centers on Kant's categorical imperative, specifically the formulation of that imperative that centers on the claim that the ultimate moral wrong is to treat someone as a means to some other end, rather than as an end in him or herself Those who do not support the use of embryos for the sole purpose of enhancing research argue that it is morally unacceptable to use embryos for scientific purposes because this is a clear use of a person as a means. Some of these same arguments can apply to the use of embryos under any circumstances. In the case of spare embryos, by contrast, many are too old or morphologically inappropriate to be implanted, and thus have no other use; it is thus argued that the use of these for research is not nearly as questionable. Moreover, opponents claim that if the cultivation of spare embryos is legalized, scientists will act on the incentive to produce as many embryos as one could produce. Even many of those who do not oppose the creation of embryos for research on Kantian grounds have voiced concern that creating embryos merely for research might cheapen the act of creation.

Clinical Implications to Assisted Reproductive Technologies Clinics

Whatever the form of embryonic stem cells to be used in research, the involvement of clinical assisted reproductive technologies (ART) embryologists, technicians, and clinicians is omnipresent. The processes whereby embryos are created (whether from donor eggs or sperm intended for research purposes, or as a by-product of reproductive health care), analyzed, stored, removed from nitrogen freezing, or destroyed are all processes that require, as a matter of course, the technologies, clinical expertise, patient population, and institutions of ART. It is thus no surprise that the largest research programs to date in the field have employed obstetricians, andrologists, reproductive endocrinologists—even ART psychologists and social workers. Ethical issues related to participation in stem cell research include three key problems. First is the question of whether and under what circumstances patients or research subjects should be allowed to participate in the donation of reproductive materials for stem cell research, particularly where that research involves the creation of embryos for research purposes. Second is the question of whether reproductive clinicians and technologists should be involved in the nonreproductive use of cloning technologies for the creation of nuclear transfer-derived stem cells. Third is whether and when clinicians involved in the derivation of embryonic stem cells should be held responsible for the failure of those cells in clinical trials or therapies using those cells. On none of these issues is there professional consensus at this point, although all three issues will receive the attention of the ethics boards of professional societies such as the American Society of Reproductive Medicine in the United States, and of bioethicists.

Human Cloning and Human Dignity: An Ethical Inquiry

The President's Council on Bioethics

Executive Summary

For the past five years, the prospect of human cloning has been the subject of considerable public attention and sharp moral debate, both in the United States and around the world. Since the announcement in February 1997 of the first successful cloning of a mammal (Dolly the sheep), several other species of mammals have been cloned. Although a cloned human child has yet to be born, and although the animal experiments have had low rates of success, the production of functioning mammalian cloned offspring suggests that the eventual cloning of humans must be considered a serious possibility.

In November 2001, American researchers claimed to have produced the first cloned human embryos, though they reportedly reached only a six-cell stage before they stopped dividing and died. In addition, several fertility specialists, both here and abroad, have announced their intention to clone human beings. The United States Congress has twice taken up the matter, in 1998 and again in 2001–2002, with the House of Representatives in July 2001 passing a strict ban on all human cloning, including the production of cloned human embryos. As of this writing, several cloning-related bills are under consideration in the Senate. Many other nations have banned human cloning, and the United Nations is considering an international convention on the subject. Finally, two major national reports have been issued on human reproductive cloning, one by the National Bioethics Advisory Commission (NBAC) in 1997, the other by the National Academy of Sciences (NAS) in January 2002. Both the NBAC and the NAS reports called for further consideration of the ethical and social questions raised by cloning.

The debate over human cloning became further complicated in 1998 when researchers were able, for the first time, to isolate human embryonic stem cells. Many scientists believe that these versatile cells, capable of becoming any type of cell in the body, hold great promise for understanding and treating many chronic diseases and conditions. Some scientists also believe that stem cells derived from cloned human embryos, produced explicitly for such research, might prove uniquely useful for studying many genetic diseases and devising novel therapies. Public reaction to the prospect of cloning-for-biomedical-research has been mixed: some Americans support it for its medical promise; others oppose it because it requires the exploitation and destruction of nascent human life, which would be created solely for research purposes.

Human Cloning: What Is at Stake?

The intense attention given to human cloning in both its potential uses, for reproduction as well as for research, strongly suggests that people do not regard it as just another new technology. Instead, we see it as something quite different, something that touches fundamental aspects of our humanity. The notion of cloning raises issues about identity and individuality, the meaning of having children, the difference between procreation and manufacture, and the relationship between the generations. It also raises new questions about the manipulation of some human beings for the benefit of others, the freedom and value of biomedical inquiry, our obligation to heal the sick (and its limits), and the respect and protection owed to nascent human life.

Finally, the legislative debates over human cloning raise large questions about the relationship between science and society, especially about whether society can or should exercise ethical and prudential control over biomedical technology and the conduct of biomedical research. Rarely has such a seemingly small innovation raised such big questions.

The Inquiry: Our Point of Departure

As Members of the President's Council on Bioethics, we have taken up the larger ethical and social inquiry called for in the NBAC and NAS reports, with the aim of advancing public understanding and informing public policy on the matter. We have attempted to consider human cloning (both for producing children and for biomedical research) within its larger human, technological, and ethical contexts, rather than to view it as an isolated technical development. We focus first on the broad human goods that it may serve as well as threaten, rather than on the immediate impact of the technique itself. By our broad approach, our starting on the plane of human goods, and our open spirit of inquiry, we hope to contribute to a richer and deeper understanding of what human cloning means, how we should think about it, and what we should do about it.

On some matters discussed in this report, Members of the Council are not of one mind. Rather than bury these differences in search of a spurious consensus, we have sought to present all views fully and fairly, while recording our agreements as well as our genuine diversity of perspectives, including our differences on the final recommendations to be made. By this means, we hope to help policymakers and the general public appreciate more thoroughly the difficulty of the issues and the competing goods that are at stake.

Fair and Accurate Terminology

There is today much confusion about the terms used to discuss human cloning, regarding both the activity involved and the entities that result. The Council stresses the importance of striving not only for accuracy but also for fairness, especially because the choice of terms can decisively affect the way questions are posed, and hence how answers are given. We have sought terminology that most accurately conveys the descriptive reality of the matter, in order that the moral arguments can then proceed on the merits. We have resisted the temptation to solve the moral questions by artful redefinition or by denying to some morally crucial element a name that makes clear that there is a moral question to be faced.

On the basis of (1) a careful analysis of the act of cloning, and its relation to the means by which it is accomplished and the purposes it may serve, and (2) an extensive critical examination of alternative terminologies, the Council has adopted the following definitions for the most important terms in the matter of human cloning:

- *Cloning:* A form of reproduction in which offspring result not from the chance union of egg and sperm (sexual reproduction) but from the deliberate replication of the genetic makeup of another single individual (asexual reproduction).

- *Human cloning:* The asexual production of a new human organism that is, at all stages of development, genetically virtually identical to a currently existing or previously existing human being. It would be accomplished by introducing the nuclear material of a human somatic cell (donor) into an oocyte (egg) whose own nucleus has been removed or inactivated, yielding a product that has a human genetic constitution virtually identical to the donor of the somatic cell. (This procedure is known as "somatic cell nuclear transfer," or SCNT). We have declined to use the terms "reproductive cloning" and "therapeutic cloning." We have chosen instead to use the following designations:

- *Cloning-to-produce-children:* Production of a cloned human embryo, formed for the (proximate) purpose of initiating a pregnancy, with the (ultimate) goal of producing a child who will be genetically virtually identical to a currently existing or previously existing individual.

- *Cloning-for-biomedical-research:* Production of a cloned human embryo, formed for the (proximate) purpose of using it in research or for extracting its stem cells, with the (ultimate) goals of gaining scientific knowledge of normal and abnormal development and of developing cures for human diseases.

- *Cloned human embryo:* (a) A human embryo resulting from the nuclear transfer process (as contrasted with a human embryo arising from the union of egg and sperm). (b) The immediate (and developing) product of the initial act of cloning, accomplished by successful SCNT, whether used subsequently in attempts to produce children or in biomedical research.

Scientific Background

Cloning research and stem cell research are being actively investigated and the state of the science is changing rapidly; significant new developments could change some of the interpretations in our report. At present, however, a few general points may be highlighted.

• *The technique of cloning.* The following steps have been used to produce live offspring in the mammalian species that have been successfully cloned. Obtain an egg cell from a female of a mammalian species. Remove its nuclear DNA, to produce an enucleated egg. Insert the nucleus of a donor adult cell into the enucleated egg, to produce a reconstructed egg. Activate the reconstructed egg with chemicals or electric current, to stimulate it to commence cell division. Sustain development of the cloned embryo to a suitable stage in vitro, and then transfer it to the uterus of a female host that has been suitably prepared to receive it. Bring to live birth a cloned animal that is genetically virtually identical (except for the mitochondrial DNA) to the animal that donated the adult cell nucleus.

• *Animal cloning:* low success rates, high morbidity. At least seven species of mammals (none of them primates) have been successfully cloned to produce live births. Yet the production of live cloned offspring is rare and the failure rate is high: more than 90 percent of attempts to initiate a clonal pregnancy do not result in successful live birth. Moreover, the live-born cloned animals suffer high rates of deformity and disability, both at birth and later on. Some biologists attribute these failures to errors or incompleteness of epigenetic reprogramming of the somatic cell nucleus.

• *Attempts at human cloning.* At this writing, it is uncertain whether anyone has attempted cloning-to-produce-children (although at least one physician is now claiming to have initiated several active clonal pregnancies, and others are reportedly working on it). We do not know whether a transferred cloned human embryo can progress all the way to live birth.

• *Stem cell research.* Human embryonic stem cells have been isolated from embryos (produced by IVF) at the blastocyst stage or from the germinal tissue of fetuses. Human adult stem (or multipotent) cells have been isolated from a variety of tissues. Such cell populations can be differentiated in vitro into a number of different cell types, and are currently being studied intensely for their possible uses in regenerative medicine. Most scientists working in the field believe that stem cells (both embryonic and adult) hold great promise as routes toward cures and treatments for many human diseases and disabilities. All stem cell research is at a very early stage, and it is too soon to tell which approaches will prove most useful, and for which diseases.

• *The transplant rejection problem.* To be effective as long-term treatments, cell transplantation therapies will have to overcome the immune rejection problem. Cells and tissues derived from adult stem cells and returned to the patient from whom they were taken would not be subject (at least in principle) to immune rejection.

• *Stem cells from cloned embryos.* Human embryonic stem cell preparations could potentially be produced by using somatic cell nuclear transfer to produce a cloned human embryo, and then taking it apart at the blastocyst stage and isolating stem cells. These stem cells would be genetically virtually identical to cells from the nucleus donor, and thus could potentially be of great value in biomedical research. Very little work of this sort has been done to date in animals, and there are as yet no published reports of cloned human embryos grown to the blastocyst stage. Although the promise of such research is at this time unknown, most researchers believe it will yield very useful and important knowledge, pointing toward new therapies and offering one of several possible routes to circumvent the immune rejection problem. Although some experimental results in animals are indeed encouraging, they also demonstrate some tendency even of cloned stem cells to stimulate an immune response.

• *The fate of embryos used in research.* All extractions of stem cells from human embryos, cloned or not, involve the destruction of these embryos.

The Ethics of Cloning-to-Produce-Children

Two separate national-level reports on human cloning (NBAC 1997; NAS 2002) concluded that attempts to clone a human being would be unethical at this time due to safety concerns and the likelihood of harm to those involved. The Council concurs in this conclusion. But we have extended the work of these distinguished bodies by undertaking a broad ethical examination of the merits of, and difficulties with, cloning-to-produce-children.

Cloning-to-produce-children might serve several purposes. It might allow infertile couples or others to have genetically-related children; permit couples at risk of conceiving a child with a genetic disease to avoid having an afflicted child; allow the bearing of a child who could become an ideal transplant donor for a particular patient in need; enable a parent to keep a living connection with a dead or dying child or spouse; or enable individuals or society to try to "replicate" individuals of great talent or beauty. These purposes have been defended by appeals to the goods of freedom, existence (as opposed to nonexistence), and well-being—all vitally important ideals.

A major weakness in these arguments supporting cloning-to-produce-children is that they overemphasize the freedom, desires, and control of parents, and pay insufficient attention to the well-being of the cloned child-to-be. The Council holds that, once the child-to-be is carefully considered, these arguments are not sufficient to overcome the powerful case against engaging in cloning-to-produce-children.

First, cloning-to-produce-children would violate the principles of the ethics of human research. Given the high rates of morbidity and mortality in the cloning of other mammals, we believe that cloning-to-produce-children would be extremely unsafe, and that attempts to produce a cloned child would be highly unethical. Indeed, our moral analysis of this matter leads us to conclude that this is not, as is sometimes implied, a merely temporary objection, easily removed by the improvement of technique. We offer reasons for believing that the safety risks might be enduring, and offer arguments in support of a strong conclusion: that conducting experiments in an effort to make cloning-to-produce-children less dangerous would itself be an unacceptable violation of the norms of research ethics. There seems to be no ethical way to try to discover whether cloning-to-produce-children can become safe, now or in the future.

If carefully considered, the concerns about safety also begin to reveal the ethical principles that should guide a broader assessment of cloning-to-produce-children: the principles of freedom, equality, and human dignity. To appreciate the broader human significance of cloning-to-produce-children, one needs first to reflect on the meaning of having children; the meaning of asexual, as opposed to sexual, reproduction; the importance of origins and genetic endowment for identity and sense of self; the meaning of exercising greater human control over the processes and "products" of human reproduction; and the difference between begetting and making. Reflecting on these topics, the Council has identified five categories of concern regarding cloning-to-produce-children. (Different Council Members give varying moral weight to these different concerns.)

• *Problems of identity and individuality.* Cloned children may experience serious problems of identity both because each will be genetically virtually identical to a human being who has already lived and because the expectations for their lives may be shadowed by constant comparisons to the life of the "original."

• *Concerns regarding manufacture.* Cloned children would be the first human beings whose entire genetic makeup is selected in advance. They might come to be considered more like products of a designed manufacturing process than "gifts" whom their parents are prepared to accept as they are. Such an attitude toward children could also contribute to increased commercialization and industrialization of human procreation.

• *The prospect of a new eugenics.* Cloning, if successful, might serve the ends of privately pursued eugenic enhancement, either by avoiding the genetic defects that may arise when human reproduction is left to chance, or by preserving and perpetuating outstanding genetic traits, including the possibility, someday in the future, of using cloning to perpetuate genetically engineered enhancements.

• *Troubled family relations.* By confounding and transgressing the natural boundaries between generations, cloning could strain the social ties between them. Fathers could become "twin brothers" to their "sons"; mothers could give birth to their genetic twins; and grandparents would also be the "genetic parents" of their grandchildren. Genetic relation to only one parent might produce special difficulties for family life.

• *Effects on society.* Cloning-to-produce-children would affect not only the direct participants but also the entire society that allows or supports this activity. Even if practiced on a small scale, it could affect the way society looks at children and set a precedent for future nontherapeutic interventions into the human genetic endowment or novel forms of control by one generation over the next. In the absence of wisdom regarding these matters, prudence dictates caution and restraint.

Conclusion: *For some or all of these reasons, the Council is in full agreement that cloning-to-produce-children is not only un-*

safe but also morally unacceptable, and ought not to be attempted.

The Ethics of Cloning-for-Biomedical-Research

Ethical assessment of cloning-for-biomedical-research is far more vexing. On the one hand, such research could lead to important knowledge about human embryological development and gene action, both normal and abnormal, ultimately resulting in treatments and cures for many dreaded illnesses and disabilities. On the other hand, the research is morally controversial because it involves the deliberate production, use, and ultimate destruction of cloned human embryos, and because the cloned embryos produced for research are no different from those that could be implanted in attempts to produce cloned children. The difficulty is compounded by what are, for now, unanswerable questions as to whether the research will in fact yield the benefits hoped for, and whether other promising and morally nonproblematic approaches might yield comparable benefits. The Council, reflecting the differences of opinion in American society, is divided regarding the ethics of research involving (cloned) embryos. *Yet we agree that all parties to the debate have concerns vital to defend, vital not only to themselves but to all of us. No human being and no society can afford to be callous to the needs of suffering humanity, or cavalier about the treatment of nascent human life, or indifferent to the social effects of adopting one course of action rather than another.*

To make clear to all what is at stake in the decision, Council Members have presented, as strongly as possible, the competing ethical cases for and against cloning-for-biomedical-research in the form of first-person attempts at moral suasion. Each case has tried to address what is owed to suffering humanity, to the human embryo, and to the broader society. Within each case, supporters of the position in question speak only for themselves, and not for the Council as a whole.

A. The Moral Case for Cloning-for-Biomedical-Research

The moral case for cloning-for-biomedical-research rests on our obligation to try to relieve human suffering, an obligation that falls most powerfully on medical practitioners and biomedical researchers. We who support cloning-for-biomedical-research all agree that it may offer uniquely useful ways of investigating and possibly treating many chronic debilitating diseases and disabilities, providing aid and relief to millions. We also believe that the moral objections to this research are outweighed by the great good that may come from it. Up to this point, we who support this research all agree. But we differ among ourselves regarding the weight of the moral objections, owing to differences about the moral status of the cloned embryo. These differences of opinion are sufficient to warrant distinguishing two different moral positions within the moral case for cloning-for-biomedical-research:

Position Number One. Most Council Members who favor cloning-for-biomedical-research do so with serious moral concerns. Speaking only for ourselves, we acknowledge the following difficulties, but think that they can be addressed by setting proper boundaries.

• *Intermediate moral status.* While we take seriously concerns about the treatment of nascent human life, we believe there are sound moral reasons for not regarding the embryo in its earliest stages as the moral equivalent of a human person. We believe the embryo has a developing and intermediate moral worth that commands our special respect, but that it is morally permissible to use early-stage cloned human embryos in important research under strict regulation.

• *Deliberate creation for use.* We believe that concerns over the problem of deliberate creation of cloned embryos for use in research have merit, but when properly understood should not preclude cloning-for-biomedical-research. These embryos would not be "created for destruction," but for use in the service of life and medicine. They would be destroyed in the service of a great good, and this should not be obscured.

• *Going too far.* We acknowledge the concern that some researchers might seek to develop cloned embryos beyond the blastocyst stage, and for those of us who believe that the cloned embryo has a developing and intermediate moral status, this is a very real worry. We approve, therefore, only of research on cloned embryos that is strictly limited to the first fourteen days of development – a point near when the primitive streak is formed and before organ differentiation occurs.

• *Other moral hazards.* We believe that concerns about the exploitation of women and about the risk that cloning-for-biomedical-research could lead to cloning-to-produce-children can be adequately addressed by appropriate rules and regulations. These concerns need not frighten us into abandoning an important avenue of research.

Position Number Two. A few Council Members who favor cloning-for-biomedical-research do not share all the ethical qualms expressed above. Speaking only for ourselves, we hold that this research, at least for the purposes presently contemplated, presents no special moral problems, and therefore should be endorsed with enthusiasm as a potential new means of gaining knowledge to serve humankind. Because we accord no special moral status to the early-stage cloned embryo and believe it should be treated essentially like all other human cells, we believe that the moral issues involved in this research are no different from those that accompany any biomedical research. What is required is the usual commitment to high standards for the quality of research, scientific integrity, and the need to obtain informed consent from donors of the eggs and somatic cells used in nuclear transfer.

B. The Moral Case against Cloning-for-Biomedical-Research

The moral case against cloning-for-biomedical-research acknowledges the possibility—though purely speculative at the moment—that medical benefits might come from this particular avenue of experimentation. But we believe it is morally wrong to exploit and destroy developing human life, even for good reasons, and that it is unwise to open the door to the many undesirable consequences that are likely to result from this research. We find it disquieting, even somewhat ignoble, to treat what are in fact seeds of the next generation as mere raw material for satisfying the needs of our own. Only for very serious reasons should progress toward increased knowledge and medical advances be slowed. But we believe that in this case such reasons are apparent.

- *Moral status of the cloned embryo.* We hold that the case for treating the early-stage embryo as simply the moral equivalent of all other human cells (Position Number Two, above) is simply mistaken: it denies the continuous history of human individuals from the embryonic to fetal to infant stages of existence; it misunderstands the meaning of potentiality; and it ignores the hazardous moral precedent that the routinized creation, use, and destruction of nascent human life would establish. We hold that the case for according the human embryo "intermediate and developing moral status" (Position Number One, above) is also unconvincing, for reasons both biological and moral. Attempts to ground the limited measure of respect owed to a maturing embryo in certain of its developmental features do not succeed, and the invoking of a "special respect" owed to nascent human life seems to have little or no operative meaning if cloned embryos may be created in bulk and used routinely with impunity. If from one perspective the view that the embryo seems to amount to little may invite a weakening of our respect, from another perspective its seeming insignificance should awaken in us a sense of shared humanity and a special obligation to protect it.

• *The exploitation of developing human life.* To engage in cloning-for-biomedical-research requires the irreversible crossing of a very significant moral boundary: the creation of human life expressly and exclusively for the purpose of its use in research, research that necessarily involves its deliberate destruction. If we permit this research to proceed, we will effectively be endorsing the complete transformation of nascent human life into nothing more than a resource or a tool. Doing so would coarsen our moral sensibilities and make us a different society: one less humble toward that which we cannot fully understand, less willing to extend the boundaries of human respect ever outward, and more willing to transgress moral boundaries once it appears to be in our own interests to do so.

• *Moral harm to society.* Even those who are uncertain about the precise moral status of the human embryo have sound ethical–prudential reasons to oppose cloning-for-biomedical-research. Giving moral approval to such research risks significant moral harm to our society by (1) crossing the boundary from sexual to asexual reproduction, thus approving in principle the genetic manipulation and control of nascent human life; (2) opening the door to other moral hazards, such as cloning-to-produce-children or research on later-stage human embryos and fetuses; and (3) potentially putting the federal government in the novel and unsavory position of mandating the destruction of nascent human life. Because we are concerned not only with the fate of the cloned embryos but also with where this research will lead our society, we think prudence requires us not to engage in this research.

• *What we owe the suffering.* We are certainly not deaf to the voices of suffering patients; after all, each of us already shares or will share in the hardships of mortal life. We and our loved ones are all patients or potential patients. But we are not only patients, and easing suffering is not our only moral obligation. As much as we wish to alleviate suffering now and to leave our children a world where suffering can be more effectively relieved, we also want to leave them a world in which we and they want to live—a world that honors moral limits, that respects all life whether strong or weak, and that refuses to secure the good of some human beings by sacrificing the lives of others.

Public Policy Options

The Council recognizes the challenges and risks of moving from moral assessment to public policy. Reflections on the "social contract" between science and society highlight both the importance of scientific freedom and the need for boundaries. We note that other countries often treat human cloning in the context of a broad area of biomedical technology, at the intersection of reproductive technology, embryo research, and genetics, while the public policy debate in the United States has treated cloning largely on its own. We recognize the special difficulty in formulating sound public policy in this area, given that the two ethically distinct matters—cloning-to-produce-children and cloning-for-biomedical-research—will be mutually affected or implicated in any attempts to legislate about either. Nevertheless, our ethical and policy analysis leads us to the conclusion that some deliberate public policy at the federal level is needed in the area of human cloning.

We reviewed the following seven possible policy options

and considered their relative strengths and weaknesses: (1) Professional self-regulation but no federal legislative action ("self-regulation"); (2) A ban on cloning-to-produce-children, with neither endorsement nor restriction of cloning-for-biomedical-research ("ban plus silence"); (3) A ban on cloning-to-produce-children, with regulation of the use of cloned embryos for biomedical research ("ban plus regulation"); (4) Governmental regulation, with no legislative prohibitions ("regulation of both"); (5) A ban on all human cloning, whether to produce children or for biomedical research ("ban on both"); (6) A ban on cloning-to-produce-children, with a moratorium or temporary ban on cloning-for-biomedical-research ("ban plus moratorium"); or (7) A moratorium or temporary ban on all human cloning, whether to produce children or for biomedical research ("moratorium on both").

The Council's Policy Recommendations

Having considered the benefits and drawbacks of each of these options, and taken into account our discussions and reflections throughout this report, the Council recommends two possible policy alternatives, each supported by a portion of the Members.

Majority Recommendation: Ten Members of the Council recommend *a ban on cloning-to-produce-children combined with a four-year moratorium on cloning-for-biomedical-research. We also call for a federal review of current and projected practices of human embryo research, pre-implantation genetic diagnosis, genetic modification of human embryos and gametes, and related matters, with a view to recommending and shaping ethically sound policies for the entire field.* Speaking only for ourselves, those of us who support this recommendation do so for some or all of the following reasons:

- By permanently banning cloning-to-produce-children, this policy gives force to the strong ethical verdict against cloning-to-produce-children, unanimous in this Council (and in Congress) and widely supported by the American people. And by enacting a four-year moratorium on the creation of cloned embryos, it establishes an additional safeguard not afforded by policies that would allow the production of cloned embryos to proceed without delay.

• It calls for and provides time for further democratic deliberation about cloning-for-biomedical research, a subject about which the nation is divided and where there remains great uncertainty. A national discourse on this subject has not yet taken place in full, and a moratorium, by making it impossible for either side to cling to the status quo, would force both to make their full case before the public. By banning all cloning for a time, it allows us to seek moral consensus on whether or not we should cross a major moral boundary (creating nascent cloned human life solely for research) and prevents our crossing it without deliberate decision. It would afford time for scientific evidence, now sorely lacking, to be gathered—from animal models and other avenues of human research—that might give us a better sense of whether cloning-for-biomedical-research would work as promised, and whether other morally nonproblematic approaches might be available. It would promote a fuller and better-informed public debate. And it would show respect for the deep moral concerns of the large number of Americans who have serious ethical objections to this research.

• Some of us hold that cloning-for-biomedical-research can never be ethically pursued, and endorse a moratorium to enable us to continue to make our case in a democratic way. Others of us support the moratorium because it would provide the time and incentive required to develop a system of national regulation that might come into use if, at the end of the four-year period, the moratorium were not reinstated or made permanent. Such a system could not be developed overnight, and therefore even those who support the research but want it regulated should see that at the very least a pause is required. In the absence of a moratorium, few proponents of the research would have much incentive to institute an effective regulatory system. Moreover, the very process of proposing such regulations would clarify the moral and prudential judgments involved in deciding whether and how to proceed with this research.

• A moratorium on cloning-for-biomedical-research would enable us to consider this activity in the larger context of research and technology in the areas of developmental biology, embryo research, and genetics, and to pursue a more comprehensive federal regulatory system for setting and executing policy in the entire area.

• Finally, we believe that a moratorium, rather than a lasting ban, signals a high regard for the value of biomedical research and an enduring concern for patients and families whose suffering such research may help alleviate. It would reaffirm the principle that science can progress while upholding the community's moral norms, and would therefore reaffirm the community's moral support for science and biomedical technology.

The decision before us is of great importance. Creating cloned embryos for *any* purpose requires crossing a major moral boundary, with grave risks and likely harms, and once we cross it there will be no turning back. Our society should take the time to make a judgment that is well informed and morally sound, respectful of strongly held views, and representative of the priorities and principles of the American people. We believe this ban-plus-moratorium proposal offers the best means of achieving these goals.

This position is supported by Council Members Rebecca S. Dresser, Francis Fukuyama, Robert P. George, Mary Ann Glendon, Alfonso Gómez-Lobo, William B. Hurlbut, Leon R. Kass, Charles Krauthammer, Paul McHugh, and Gilbert C. Meilaender.

Minority Recommendation: Seven Members of the Council recommend *a ban on cloning-to-produce-children, with regulation of the use of cloned embryos for biomedical research.* Speaking only for ourselves, those of us who support this recommendation do so for some or all of the following reasons:

• By permanently banning cloning-to-produce-children, this policy gives force to the strong ethical verdict against cloning-to-produce-children, unanimous in this Council (and in Congress) and widely supported by the American people. We believe that a ban on the transfer of cloned embryos to a woman's uterus would be a sufficient and effective legal safeguard against the practice.

• *It approves cloning-for-biomedical-research and permits it to proceed without substantial delay.* This is the most important advantage of this proposal. The research shows great promise, and its actual value can only be determined by allowing it to go forward now. Regardless of how much time we allow it, no amount of experimentation with animal models can provide the needed understanding of human diseases. The special benefits from working with stem cells from cloned human embryos cannot be obtained using embryos obtained by IVF. We believe this research could provide relief to millions of Americans, and that the government should therefore support it, within sensible limits imposed by regulation.

• It would establish, *as a condition of proceeding*, the necessary regulatory protections to avoid abuses and misuses of cloned embryos. These regulations might touch on the secure handling of embryos, licensing and prior review of research projects, the protection of egg donors, and the provision of equal access to benefits.

• Some of us also believe that mechanisms to regulate cloning-for-biomedical-research should be part of a larger regulatory program governing all research involving human embryos, and that the federal government should initiate a review of present and projected practices of human embryo research, with the aim of establishing reasonable policies on the matter.

Permitting cloning-for-biomedical-research now, while governing it through a prudent and sensible regulatory regime, is the most appropriate way to allow important research to proceed while insuring that abuses are prevented. We believe that the legitimate concerns about human cloning expressed throughout this report are sufficiently addressed by this ban-plus-regulation proposal, and that the nation should affirm and support the responsible effort to find treatments and cures that might help many who are suffering.

This position is supported by Council Members Elizabeth H. Blackburn, Daniel W. Foster, Michael S. Gazzaniga, William F. May, Janet D. Rowley, Michael J. Sandel, and James Q. Wilson.

Crossing Lines

Charles Krauthammer

The Problem

You were once a single cell. Every one of the 100 trillion cells in your body today is a direct descendent of that zygote, the primordial cell formed by the union of mother's egg and father's sperm. Each one is genetically identical (allowing for copying errors and environmental damage along the way) to that cell. Therefore, if we scraped a cell from, say, the inner lining of your cheek, its DNA would be the same DNA that, years ago in the original zygote, contained the entire plan for creating you and every part of you.

Here is the mystery: Why can the zygote, as it multiplies, produce every different kind of cell in the body—kidney, liver, brain, skin—while the skin cell is destined, however many times it multiplies, to remain skin forever? As the embryo matures, cells become specialized and lose their flexibility and plasticity. Once an adult cell has specialized-differentiated, in scientific lingo—it is stuck forever in that specialty. Skin is skin; kidney is kidney.

Understanding that mystery holds the keys to the kingdom. The Holy Grail of modern biology is regenerative medicine. If we can figure out how to make a specialized adult cell dedifferentiate—unspecialize, i.e., revert way back to the embryonic stage, perhaps even to the original zygotic stage—and then grow it like an embryo under controlled circumstances, we could reproduce for you every kind of tissue or organ you might need. We could create a storehouse of repair parts for your body. And, if we let that dedifferentiated cell develop completely in a woman's uterus, we will have created a copy of you, your clone.

That is the promise and the menace of cloning. It has already been done in sheep, mice, goats, pigs, cows, and now cats and rabbits (though cloning rabbits seems an exercise in biological redundancy). There is no reason in principle why it cannot be done in humans. The question is: Should it be done?

Notice that the cloning question is really two questions: (1) May we grow that dedifferentiated cell all the way into a cloned baby, a copy of you? That is called reproductive cloning. And (2) may we grow that dedifferentiated cell just into the embryonic stage and then mine it for parts, such as stem cells? That is called research cloning.

Reproductive cloning is universally abhorred. In July 2001 the House of Representatives, a fairly good representative of the American people, took up the issue and not a single member defended reproductive cloning. Research cloning, however, is the hard one. Some members were prepared to permit the cloning of the human embryo in order to study and use its component parts, with the proviso that the embryo be destroyed before it grows into a fetus or child. They were a minority, however. Their amendment banning baby-making but permitting research cloning was defeated by 76 votes. On July 31, 2001, a bill outlawing all cloning passed the House decisively.

Within weeks, perhaps days, the Senate will vote on essentially the same alternatives. On this vote will hinge the course of the genetic revolution at whose threshold we now stand.

The Promise

This is how research cloning works. You take a donor egg from a woman, remove its nucleus, and inject the nucleus of, say, a skin cell from another person. It has been shown in animals that by the right manipulation you can trick the egg and the injected nucleus into dedifferentiating—that means giving up all the specialization of the skin cell and returning to its original state as a primordial cell that could become anything in the body.

In other words, this cell becomes totipotent. It becomes the equivalent of the fertilized egg in normal procreation, except that instead of having chromosomes from two people, it has chromosomes from one. This cell then behaves precisely like an embryo. It divides. It develops. At four to seven days, it forms a "blastocyst" consisting of about 100 to 200 cells.

The main objective of cloning researchers would be to disassemble this blastocyst: pull the stem cells out, grow them in the laboratory, and then try to tease them into becoming specific kinds of cells, say, kidney or heart or brain and so on.

There would be two purposes for doing this: study or cure. You could take a cell from a person with a baffling disease, like Lou Gehrig's, clone it into a blastocyst, pull the stem cells out, and then study them in order to try to understand the biology of the illness. Or you could begin with a cell from a person with Parkinson's or a spinal cord injury, clone it, and tease out the stem cells to develop tissue that you would reinject into the original donor to, in theory, cure the Parkinson's or spinal cord injury. The advantage of using a cloned cell rather than an ordinary stem cell is that, presumably, there would be no tissue rejection. It's your own DNA. The body would recognize it. You'd have a perfect match.

(Research cloning is sometimes called therapeutic cloning, but that is a misleading term. First, because therapy by reinjection is only one of the many uses to which this cloning can be put. Moreover, it is not therapeutic for the clone—indeed, the clone is invariably destroyed in the process—though it may be therapeutic for others. If you donate a kidney to your brother, it would be odd to call your operation a therapeutic nephrectomy. It is not. It's a sacrificial nephrectomy.)

The conquest of rejection is one of the principal rationales for research cloning. But there is reason to doubt this claim on scientific grounds. There is some empirical evidence in mice that cloned tissue may be rejected anyway (possibly because a clone contains a small amount of foreign- mitochondrial—DNA derived from the egg into which it was originally injected). Moreover, enormous advances are being made elsewhere in combating tissue rejection. The science of immune rejection is much more mature than the science of cloning. By the time we figure out how to do safe and reliable research cloning, the rejection problem may well be solved. And finally, there are less problematic alternatives—such as adult stem cells—that offer a promising alternative to cloning because they present no problem of tissue rejection and raise none of cloning's moral conundrums.

These scientific considerations raise serious questions about the efficacy of, and thus the need for, research cloning. But there is a stronger case to be made. Even if the scientific objections are swept aside, even if research cloning is as doable and promising as its advocates contend, there are other reasons to pause.

The most obvious is this: Research cloning is an open

door to reproductive cloning. Banning the production of cloned babies while permitting the production of cloned embryos makes no sense. If you have factories all around the country producing embryos for research and commerce, it is inevitable that someone will implant one in a woman (or perhaps in some artificial medium in the farther future) and produce a human clone. What then? A law banning reproductive cloning but permitting research cloning would then make it a crime not to destroy that fetus—an obvious moral absurdity.

This is an irrefutable point and the reason that many in Congress will vote for the total ban on cloning. Philosophically, however, it is a showstopper. It lets us off too early and too easy. It keeps us from facing the deeper question: Is there anything about research cloning that in and of itself makes it morally problematic?

Objection I: Intrinsic Worth

For some people, life begins at conception. And not just life—if life is understood to mean a biologically functioning organism, even a single cell is obviously alive—but personhood. If the first zygotic cell is owed all the legal and moral respect due a person, then there is nothing to talk about. Ensoulment starts with Day One and Cell One, and the idea of taking that cell or its successor cells apart to serve someone else's needs is abhorrent.

This is an argument of great moral force but little intellectual interest. Not because it may not be right. But because it is unprovable. It rests on metaphysics. Either you believe it or you don't. The discussion ends there.

I happen not to share this view. I do not believe personhood begins at conception. I do not believe a single cell has the moral or legal standing of a child. This is not to say that I do not stand in awe of the developing embryo, a creation of majestic beauty and mystery. But I stand in equal awe of the Grand Canyon, the spider's web, and quantum mechanics. Awe commands wonder, humility, appreciation. It does not command inviolability. I am quite prepared to shatter an atom, take down a spider's web, or dam a canyon for electricity. (Though we'd have to be very short on electricity before I'd dam the Grand.)

I do not believe the embryo is entitled to inviolability. But is it entitled to nothing? There is a great distance between inviolability, on the one hand, and mere "thingness," on the other. Many advocates of research cloning see nothing but thingness. That view justifies the most ruthless exploitation of the embryo. That view is dangerous.

Why? Three possible reasons. First, the Brave New World Factor: Research cloning gives man too much power for evil. Second, the Slippery Slope: The habit of embryonic violation is in and of itself dangerous. Violate the blastocyst today and every day, and the practice will inure you to violating the fetus or even the infant tomorrow. Third, Manufacture: The very act of creating embryos for the sole purpose of exploiting and then destroying them will ultimately predispose us to a ruthless utilitarianism about human life itself.

Objection II: The Brave New World Factor

The physicists at Los Alamos did not hesitate to penetrate, manipulate, and split uranium atoms on the grounds that uranium atoms possess intrinsic worth that entitled them to inviolability. Yet after the war, many fought to curtail atomic power. They feared the consequences of delivering such unfathomable power—and potential evil—into the hands of fallible human beings. Analogously, one could believe that the cloned blastocyst has little more intrinsic worth than the uranium atom and still be deeply troubled by the manipulation of the blastocyst because of the fearsome power it confers upon humankind.

The issue is leverage. Our knowledge of how to manipulate human genetics (or atomic nuclei) is still primitive. We could never construct ex nihilo a human embryo. It is an unfolding organism of unimaginable complexity that took nature three billion years to produce. It might take us less time to build it from scratch, but not much less. By that time, we as a species might have acquired enough wisdom to use it wisely. Instead, the human race in its infancy has stumbled upon a genie infinitely too complicated to create or even fully understand, but understandable enough to command and perhaps even control. And given our demonstrated unwisdom with our other great discovery—atomic power: As we speak, the very worst of humanity is on the threshold of acquiring the most powerful weapons in history—this is a fear and a consideration to be taken very seriously.

For example. Female human eggs seriously limit the mass production of cloned embryos. Extracting eggs from women is difficult, expensive, and potentially dangerous. The search is on, therefore, for a good alternative. Scientists have begun injecting human nuclei into the egg cells of animals. In 1996 Massachusetts scientists injected a human nucleus with a cow egg. Chinese scientists have fused a human fibroblast with a rabbit egg and have grown the resulting embryo to the blastocyst stage. We have no idea what grotesque results might come from such interspecies clonal experiments.

In October 2000 the first primate containing genes from another species was born (a monkey with a jellyfish gene). In 1995 researchers in Texas produced headless mice. In 1997 researchers in Britain produced headless tadpoles. In theory, headlessness might be useful for organ transplantation. One can envision, in a world in which embryos are routinely manufactured, the production of headless clones—subhuman creatures with usable human organs but no head, no brain, no consciousness to identify them with the human family.

The heart of the problem is this: Nature, through endless evolution, has produced cells with totipotent power. We are about to harness that power for crude human purposes. That should give us pause. Just around the corner lies the logical by-product of such power: human–animal hybrids, partly developed human bodies for use as parts, and other horrors imagined—Huxley's Deltas and Epsilons—and as yet unimagined. This is the Brave New World Factor. Its grounds for objecting to this research are not about the beginnings of life, but about the ends; not the origin of these cells, but their destiny; not where we took these magnificent cells from, but where they are taking us.

Objection III: The Slippery Slope

The other prudential argument is that once you start tearing apart blastocysts, you get used to tearing apart blastocysts. And whereas now you'd only be doing that at the seven-day stage, when most people would look at this tiny clump of cells on the head of a pin and say it is not inviolable, it is inevitable that some scientist will soon say: Give me just a few more weeks to work with it and I could do wonders.

That will require quite a technological leap because the blastocyst will not develop as a human organism unless implanted in the uterus. That means that to go beyond that seven-day stage you'd have to implant this human embryo either in an animal uterus or in some fully artificial womb.

Both possibilities may be remote, but they are real. And then we'll have a scientist saying: Give me just a few more months with this embryo, and I'll have actual kidney cells, brain cells, pancreatic cells that I can transplant back into the donor of the clone and cure him. Scientists at Advanced Cell Technology in Massachusetts have already gone past that stage in animals. They have taken cloned cow embryos past the blastocyst stage, taken tissue from the more developed cow fetus, and reimplanted it back into the donor animal.

The scientists' plea to do the same in humans will be hard to ignore. Why grow the clone just to the blastocyst stage, destroy it, pull out the inner cell mass, grow stem cells out of that, propagate them in the laboratory, and then try chemically or otherwise to tweak them into becoming kidney cells or brain cells or islet cells? This is Rube Goldberg. Why not just allow that beautiful embryonic machine, created by nature and far more sophisticated than our crude techniques, to develop unmolested? Why not let the blastocyst grow into a fetus that possesses the kinds of differentiated tissue that we could then use for curing the donor?

Scientifically, this would make sense. Morally, we will have crossed the line between tearing apart a mere clump of cells and tearing apart a recognizable human fetus. And at that point, it would be an even smaller step to begin carving up seven- and eight-month-old fetuses with more perfectly formed organs to alleviate even more pain and suffering among the living. We will, slowly and by increments, have gone from stem cells to embryo farms to factories with fetuses in various stages of development and humanness, hanging (metaphorically) on meat hooks waiting to be cut open to be used by the already born.

We would all be revolted if a living infant or developed fetus were carved up for parts. Should we build a fence around that possibility by prohibiting any research on even the very earliest embryonic clump of cells? Is the only way to avoid the slide never to mount the slippery slope at all? On this question, I am personally agnostic. If I were utterly convinced that we would never cross the seven-day line, then I would have no objection on these grounds to such research on the inner cell mass of a blastocyst. The question is: Can we be sure? This is not a question of principle; it is a question of prudence. It is almost a question of psychological probability. No one yet knows the answer.

Objection IV: Manufacture

Note that while, up to now, I have been considering arguments against research cloning, they are all equally applicable to embryonic research done on a normal—i.e., noncloned—embryo. If the question istearing up the blastocyst, there is no intrinsic moral difference between a two-parented embryo derived from a sperm and an egg and a single-parented embryo derived from a cloned cell. Thus the various arguments against this research—the intrinsic worth of the embryo, the prudential consideration that we might create monsters, or the prudential consideration that we might become monsters in exploiting post-embryonic forms of human life (fetuses or even children)—are identical to the arguments for and against stem-cell research.

These arguments are serious—serious enough to banish the insouciance of the scientists who consider anyone questioning their work to be a Luddite—yet, in my view, insufficient to justify a legal ban on stem-cell research (as with stem cells from discarded embryos in fertility clinics). I happen not to believe that either personhood or ensoulment occurs at conception. I think we need to be apprehensive about what evil might arise from the power of stem-cell research, but that apprehension alone, while justifying vigilance and regulation, does not justify a ban on the practice. And I believe that given the good that might flow from stem-cell research, we should first test the power of law and custom to enforce the seven-day blastocyst line for embryonic exploitation before assuming that such a line could never hold.

This is why I support stem-cell research (using leftover embryos from fertility clinics) and might support research cloning were it not for one other aspect that is unique to it. In research cloning, the embryo is created with the explicit intention of its eventual destruction. That is a given because not to destroy the embryo would be to produce a cloned child. If you are not permitted to grow the embryo into a child, you are obliged at some point to destroy it.

Deliberately creating embryos for eventual and certain destruction means the launching of an entire industry of embryo manufacture. It means the routinization, the commercialization, the commodification of the human embryo. The bill that would legalize research cloning essentially sanctions, licenses, and protects the establishment of a most ghoulish enterprise: the creation of nascent human life for the sole purpose of its exploitation and destruction.

How is this morally different from simply using discarded embryos from in vitro fertilization (IVF) clinics? Some have suggested that it is not, that to oppose research cloning is to oppose IVF and any stem-cell research that comes out of IVF. The claim is made that because in IVF there is a high probability of destruction of the embryo, it is morally equivalent to research cloning. But this is plainly not so. In research cloning there is not a high probability of destruction; there is 100 percent probability. Because every cloned embryo must be destroyed, it is nothing more than a means to someone else's end.

In IVF, the probability of destruction may be high, but it need not necessarily be. You could have a clinic that produces only a small number of embryos, and we know of many cases of multiple births resulting from multiple embryo implantation. In principle, one could have IVF using only a single embryo and thus involving no deliberate embryo destruction at all. In principle, that is impossible in research cloning.

Furthermore, a cloned embryo is created to be destroyed and used by others. An IVF embryo is created to develop into a child. One cannot disregard intent in determining morality. Embryos are created in IVF to serve reproduction. Embryos are created in research cloning to serve, well, research. If certain IVF embryos were designated as "helper embryos" that would simply aid an anointed embryo in turning into a child, then we would have an analogy to cloning. But, in fact, we don't know which embryo is anointed in IVF. They are all created to have a chance of survival. And they are all equally considered an end.

Critics counter that this ends-and-means argument is really obfuscation, that both procedures make an instrument of the embryo. In cloning, the creation and destruction of the embryo is a means to understanding or curing disease. In IVF, the creation of the embryo is a means of satisfying a couple's need for a child. They are both just means to ends.

But it makes no sense to call an embryo a means to the creation of a child. The creation of a child is the destiny of an embryo. To speak of an embryo as a means to creating a child empties the word "means" of content. The embryo in IVF is a stage in the development of a child; it is no more a means than a teenager is a means to the adult he or she later becomes. In contrast, an embryo in research cloning is pure means. Laboratory pure.

And that is where we must draw the line. During the great debate on stem-cell research, a rather broad consensus was reached (among those not committed to "intrinsic worth" rendering all embryos inviolable) that stem-cell research could be morally justified because the embryos destroyed for their possibly curative stem cells were derived from fertility clinics and thus were going to be discarded anyway. It was understood that human embryos should not be created solely for the purpose of being dismembered and then destroyed for the benefit of others. Indeed, when Senator Bill Frist made his impassioned presentation on the floor of the Senate supporting stem-cell research, he included among his conditions a total ban on creating human embryos just to be stem-cell farms.

Where cloning for research takes us decisively beyond stem-cell research is in sanctioning the manufacture of the human embryo. You can try to regulate embryonic research to prohibit the creation of Brave New World monsters; you can build fences on the slippery slope, regulating how many days you may grow an embryo for research; but once you countenance the very creation of human embryos for no other purpose than for their parts, you have crossed a moral frontier.

Research cloning is the ultimate in conferring thingness up on the human embryo. It is the ultimate in desensitization. And as such, it threatens whatever other fences and safeguards we might erect around embryonic research. The problem, one could almost say, is not what cloning does to the embryo, but what it does to us. Except that, once cloning has changed us, it will inevitably enable further assaults on human dignity. Creating a human embryo just so it can be used and then destroyed undermines the very foundation of the moral prudence that informs the entire enterprise of genetic research: the idea that, while a human embryo may not be a person, it is not nothing. Because if it is nothing, then everything is permitted. And if everything is permitted, then there are no fences, no safeguards, no bottom.

Would Cloned Humans Really Be Like Sheep?

Leon Eisenberg

The recent proof, by DNA-microsatellite analysis[1] and DNA-fingerprinting techniques,[2] that Dolly the sheep had indeed been cloned as Wilmut et al. claimed,[3] and the report by Wakayama et al.[4] of the successful cloning of more than 20 healthy female mice are likely to reactivate discussions of the ethics of cloning humans and to provoke more calls to ban experiments on mammalian cloning altogether. From the standpoint of biologic science, a ban on such laboratory experiments would be a severe setback to research in embryology.[5] From the standpoint of moral philosophy, the ethical debate has been so obscured by incorrect assumptions about the relation between a potential human clone and its adult progenitor that the scientific issues must be reexamined in order to clarify the relation between genotype and phenotype. There are powerful biologic objections to the use of cloning to alter the human species, objections that make speculations about the ethics of the process largely irrelevant.

Experiments in Cloning

A clone is the aggregate of the asexually produced progeny of an individual organism. Reproduction by cloning in horticulture involves the use of cuttings of a single plant to propagate desired botanical characteristics indefinitely. In microbiology, a colony of bacteria constitutes a clone if its members are the descendants of a single bacterium that has undergone repeated asexual fission. The myriad bacteria in the clone each have precisely the same genetic complement as that of the progenitor cell and are indistinguishable from one another.

Success in cloning mammals demonstrates unequivocally that at least some of the nuclei in fully differentiated mammalian cells contain the full complement of potentially active genetic material that is present in the zygote. What distinguishes differentiated cells is the sets of genes that are turned "off" or "on." The cloning experiments in animals suggest that similar techniques might make it possible to clone humans. Such cloning would involve transferring a human ovum to a test tube, removing its nucleus, replacing it with a somatic-cell nucleus from the donor of the ovum or another person, allowing the ovum with its new diploid nucleus to differentiate to the blastula stage, and then implanting it in a "host" uterus. The resultant person, on attaining maturity, would be an identical genetic twin of the adult nuclear donor. This hypothetical outcome, although remote, has given rise to speculation about the psychological, ethical, and social consequences of producing clones of human beings. The futuristic scenarios evoked by the prospect of human cloning contain implicit assumptions about the mechanisms of human development. Examination of these underlying premises highlights themes that can be traced back to Greek antiquity, themes that recur in contemporary debates about the sources of differences between groups with respect to such characteristics as intelligence and aggression.[6]

Theories of Development

The enigmas of human development have concerned philosophers and naturalists since people first began to wonder how plants and animals emerged from the products of fertilization.[7] Despite the fact that there is no resemblance between the physical appearance of the seed and the form of the adult organism, the plant or animal to which it gives rise is an approximate replica of its progenitors. The earliest Greek explanation was preformation—that is, the seed contains all adult structures in miniature. This ancient speculation, found in the Hippocratic corpus, was given poetic expression by Seneca:[8] "In the seed are enclosed all the parts of the body of the man that shall be formed. The infant that is born in his mother's womb has the roots of the beard and hair that he shall wear one day." The theory of preformation was so powerful that 1600 years later, when the microscope was invented, the first microscopists to examine a sperm were able to persuade themselves that they could see in its head a homunculus with all the features of a tiny but complete man. Improvements in the microscope and the establishment of embryology as an experimental science made the doctrine progressively more difficult to sustain in its original form. With better microscopical resolution, the expected structures could not be seen, and experimental manipulation of embryos produced abnormal "monsters" that' could not, by definition, have already been present in the seed.

The alternative view, that of epigenesis, was formulated by Aristotle. Having opened eggs at various stages of development, he observed that the individual organs did not all appear at the same time, as preformation theory demanded. He did not accept the argument that differences in the size of the organs could account for the failure to see them all at the same time. Others as well as Aristotle had noted that the heart is visible before the lungs, even though the lungs are ultimately much larger. Unlike his predecessors, Aristotle began with the observable data. He concluded that new parts were formed in succession and did not merely unfold from precursors already present: [9]

> It is possible, then, that A should move B and B should move C, that, in fact, the case should be the same as with the automatic machines shown as curiosities. For the parts of such machines while at rest have a sort of potentiality of motion in them, and when any external force puts the first of them into motion, immediately the next is moved in actuality... In like manner also that from which the semen comes...sets up the movement in the embryo and makes the parts of it by having touched first something though not continuing to touch it... While there is something that makes the parts, this does not exist as a definite object, nor does it exist in the semen at the first as a complete part.

This is the first statement of the theory of epigenesis: successive stages of differentiation in the course of development give rise to new properties and new structures. The genetic code in the zygote determines the range of possible outcomes. Yet the genes that are active in the zygote serve only to initiate a sequence, the outcome of which is dependent on the moment-to-moment interactions between the products of successive stages in development. For example, the potential for differentiating into pancreatic tissue is limited to cells in a particular zone of the embryo. But these cells will produce prozymogen, the histologic marker of pancreatic tissue, only if they are in contact with neighboring mesenchymal cells; if they are separated from mesenchyme, their evolution is arrested, despite their genetic potential.[10] At the same time, the entire process is dependent on the adequacy of the uterine environment, defects in which lead to anomalous development and miscarriage.

Outcomes of Human Cloning

The methodological barriers to successful human cloning are formidable. Nonetheless, even if the necessary virtuosity lies in a more distant future than science-fiction enthusiasts suggest, one can argue that a solution exists in principle and attempt to envisage the possible outcomes.

Restricting Genetic Diversity

One negative consequence of very wide scale cloning is that it would lead to a marked restriction in the diversity of the human gene pool. Such a limitation would endanger the ability of our species to survive major environmental changes. Genetic homogeneity is compatible only with adaptation to a very narrow ecological niche. Once that niche is perturbed (e.g., by the invasion of a new predator or a change in temperature or water supply), extinction may follow. For example, the "green revolution" in agriculture has led to the selective cultivation of grain seeds chosen for high yield under modern conditions of fertilization and pest control. Worldwide food production, as a result, is now highly vulnerable to new blights because of our reliance on a narrow range of genotypes.[11] Recognition of this threat has led to a call for the creation of seed banks containing representatives of "wild" species as protection against catastrophe from new blights or changed climatic conditions, to which the current high-yield grains prove particularly vulnerable.[12] Indeed, the loss of species (genetically distinct populations) is impoverishing global biodiversity as the result of shrinking habitats.[13]

Precisely the same threat would hold for humans, were we to replace sexual reproduction with cloning. The extraordinary biologic investment in sexual reproduction (as compared with asexual replication) provides a measure of its importance to the evolution of species. Courtship is expensive in its energy requirements, reproductive organs are elaborate, and there are extensive differences between male and female in secondary sexual characteristics. The benefit of sexual reproduction is the enhancement of diversity (by the crossover between homologous chromosomes during meiosis and by the combining of the haploid gametes of a male and a female). The new genetic combinations so produced enable the species to respond as a population to changing environmental conditions through the selective survival of adaptable genotypes.

Cloning Yesterday's People for Tomorrow's Problems

The choice of whom to clone could be made only on the basis of phenotypic characteristics manifested during the several decades when the persons being considered for cloning had come to maturity. Let us set aside the problem of assigning value to particular characteristics and assume that we agree on the traits to be valued, however unrealistic that assumption.

By definition, the genetic potential for these characteristics must have existed in the persons who now exhibit them. But the translation of that potential into the phenotype occurs in the particular environment in which development occurs. Even if we agree on the genotype we wish to preserve, we face a formidable barrier: we know so little of the environmental features necessary for the flowering of that genotype that we cannot specify in detail the environment we would have to provide, both before and after birth, to ensure a phenotypic outcome identical to the complex of traits we seek to perpetuate.

Let us make a further dubious assumption and suppose the day has arrived when we can specify the environment necessary for the flowering of the chosen phenotype. Nonetheless, the phenotype so admirably suited to the world in which it matured may not be adaptive to the world a generation hence. That is, the traits that lead a person to be creative or to exhibit leadership at one moment in history may not be appropriate at another. Not only is the environment not static, it is altered by our own extraordinary impact on our ecology. The proliferation of our species changes patterns of disease[14, 15]; our methods of disease control, by altering population ratios, affect the physical environment itself.[16] Social evolution demands new types of men and women. Cloning would condemn us always to plan the future on the basis of the past (since the successful phenotype cannot be identified sooner than adulthood).

The Connection between Genotype and Phenotype

For the student of biology, cloning is a powerful and instructive method with great potential for deepening our understanding of the mechanisms of differentiation during development. The potential of a given genotype can only be estimated from the varied manifestations of the phenotype over as wide a range of environments as are compatible with its survival. The wider the range of environments, the greater the diversity observed in the phenotypic manifestations of the one genotype. Human populations possess an extraordinary range of latent variability. Dissimilar genotypes can produce remarkably similar phenotypes under the wide range of conditions that characterize the environments of the inhabitable portions of the globe. The differences resulting from genotypic variability are manifested most clearly under extreme conditions, when severe stresses overwhelm the homeostatic mechanisms that ordinarily act as buffers against small perturbations.

Phenotypic identity requires identity between genotypes, which cloning can ensure, and identity between environmental interactions, which it cannot ensure. At the most trivial level, we can anticipate less similarity even in physical appearance between cell donor and cloned recipient than that which is observed between one-egg twins. Placental attachment and fetal-maternal circulation can vary substantially, even for uniovular twins housed in one uterus. Developmental circumstances will be more variable between donor and cloned recipient, who will have been carried by different women.

Postnatal Environmental Effects on the Human Brain

Let us force the argument one step further by assuming that the environmental conditions for the cloned infant have been identical to those of his or her progenitor, so that at birth the infant is a replica of the infant its "father" or "mother" was at birth. Under such circumstances (and within the limits of the precision of genetic specification), the immediate pattern of central nervous system connections and their responses to stimulation will be the same as those of the progenitor at birth.

However, even under these circumstances, the future is not predestined. The human species is notable for the proportion of brain development that occurs postnatally. Other primate brains increase in weight from birth to maturity by a factor of 2 to 2.5, but the human brain increases by a factor of 3.5 to 4.

There is a fourfold increase in the neocortex, with a marked elaboration of the receiving areas for the teloreceptors, a disproportionate expansion of the motor area for the hand in relation to the representation of other parts, a representation of tongue and larynx many times greater, and a great increase in the "association" areas. The elaboration of pathways and interconnections is highly dependent on the quantity, quality, and timing of intellectual and emotional stimulation. The very structure of the brain, as well as the function of the mind, emerges from the interaction between maturation and experience.[17]

Nature and nurture jointly mold the structure of the brain. The basic plan of the central nervous system is laid down in the human genome, but the detailed pattern of connections results from competition between axons for common target neurons. Consider the steps in the formation of alternating ocular layers in the lateral geniculate bodies. Early in embryogenesis, axons from both eyes enter each of the geniculate nuclei and intermingle. How does the separation of layers for each eye, essential for vision, come about? It results from periodic waves of spontaneous electrical activity in retinal ganglion cells, because immature cell membranes are unstable. If these electrical outbursts are abolished experimentally, the layers simply do not become separated.[18] Competition between the two eyes, driven by spontaneous retinal activity, determines eye-specific lateral geniculate connections.[19] Neither the genes governing the retina nor the genes governing the geniculate specify the alternating ocular layers; it is the interaction between retina and geniculate during embryogenesis that brings it about. Furthermore, the precise targeting of projections from lateral geniculate to occipital cortex is dependent on electrical activity in the geniculate. Abolishing these action potentials with an infusion of tetrodotoxin results in projections to cortical areas that are normally bypassed and a marked reduction in projections to visual cortex.[20]

Postnatal stimulation is required to form the ocular dominance columns in the occipital cortex.[21] Both eyes of the newborn must receive precisely focused stimulation from the visual environment during the early months of postnatal life in order to fine-tune the structure of the cortex. If focused vision in one eye of a kitten or an infant monkey is interfered with, the normal eye "captures" most neurons in the occipital cortex in the absence of competition from the deprived eye. The change becomes irreversible if occlusion is maintained throughout the sensitive period. Amblyopia in humans, characterized by incongruent visual images from the two eyes, results in permanent loss of effective vision from the unused eye if the defect is not corrected within the first five years of life.

Thus, which of the overabundant neurons live and which

die is determined by the amount and consistency of the stimulation they receive. Interaction between organism and environment leads to patterned neuronal activity that determines which synapses will persist.[22] Experience molds the brain in a process that continues throughout life. Myelination in a key relay zone in the hippocampal formation continues to increase from childhood through at least the sixth decade of life.[23] And recent research has provided evidence that neurons in the dentate gyrus of the hippocampus continue to divide in the adult brain.[24]

Changes in the Brain with Use

Techniques of functional brain mapping reveal marked variations in cortical representation that depend on prior experience. Manipulation of sensory inputs leads to reorganization of the cortex in monkey[25] and humans.[26] The motor cortex in violinists displays a substantially larger representation of the fingers of the left hand (the one used to play the strings) than of the fingers of the right (or bowing) hand. Moreover, the area of the brain dedicated to finger representation is larger in musicians than in nonmusicians.[27] Sterr et al[28] compared finger representation in the somatosensory cortex in blind persons who used three fingers on each hand to read Braille with that in Braille readers using only one finger on one hand and in sighted readers. They found a substantial enlargement of hand representation in the Braille readers who used two hands, with topographic changes on the postcentral gyrus.

If enlargement of cortical areas accompanies increases in activity, shrinkage follows loss. Within days after mastectomy, the amputation of an arm or leg, or the correction of syndactyly, the cortical sensory map changes. Intact areas have an enlarged representation at the expense of areas from which innervation has been removed.[29, 30] What begins prenatally continues throughout life. Structure follows function.

Becoming Human

There is yet another level of complexity in the analysis of personality development. The human traits of interest to us are polygenic rather than monogenic; similar outcomes can result from the interaction between different genomes and different social environments. To produce another Wolfgang Amadeus Mozart, we would need not only Wolfgang's genome but his mother's uterus, his father's music lessons, his parents' friends and his own, the state of music in 18th-century Austria, Haydn's patronage, and on and on, in ever-widening circles. Without Mozart's set of genes, the rest would not suffice; there was, after all, only one Wolfgang Amadeus Mozart. But we have no right to the converse assumption: that his genome, cultivated in another world at another time, would result in the same musical genius. If a particular strain of wheat yields different harvests under different conditions of climate, soil, and cultivation, how can we assume that so much more complex a genome as that of a human being would yield its desired crop of operas, symphonies, and chamber music under different circumstances of nurture?

In sum, cloning would be a poor method indeed for improving on the human species. If widely adopted, it would have a devastating impact on the diversity of the human gene pool. Cloning would select for traits that have been successful in the past but that will not necessarily be adaptive to an unpredictable future. Whatever phenotypes might be produced would be extremely vulnerable to the uncontrollable vicissitudes of the environment.

Proposals for human cloning as a method for "improving" the species are biologic nonsense. To elevate the question to the level of an ethical issue is sheer casuistry. The problem lies not in the ethics of cloning a human but in the metaphysical cloud that surrounds this hypothetical cloned creature. Pseudobiology trivializes ethics and distracts our attention from real moral issues: the ways in which the genetic potential of humans born into impoverished environments today is stunted and thwarted. To improve our species, no biologic sleight of hand is needed. Had we the moral commitment to provide every child with what we desire for our own, what a flowering of humankind there would be.

References

1. Ashworth D, Bishop M, Campbell K, et al. DNA microsatellite analysis of Dolly. Nature 1998;394:329.
2. Signer EN, Dubrova YE, Jeffreys AJ, et al. DNA fingerprinting Dolly Nature 1998;394:329–30.
3. Wilmut I, Schnicke AE, McWhir J, Kind AJ, Campbell KH. Viable off spring derived from fetal and adult mammalian cells. Nature 1997;385: 810–3. [Erratum, Nature 1997;386:200.]
4. Wakayama T, Perry AC, Zuccotti M, Johnson KR, Yanagimachi R. Full-term development of mice from enucleated oocytes injected with cumulus cell nuclei. Nature 1998;394:369–74.
5. Berg P, Singer M. Regulating human cloning. Science 1998;282:413.
6. Eisenberg L. The human nature of human nature. Science 1972;176:123–8.
7. Needham J. A history of embryology. New York: Abelard-Schuman, 1959.
8. Id., A history of embryology. New York: Abelard-Schuman, 1959:66.
9. Id. A history of embryology. New York: Abelard-Schuman, 1959:47–8.
10. Grobstein C. Cytodifferentiation and its controls. Science 1964;143:643–50.
11. Harlan JR. Our vanishing genetic resources. Science 1975;188:618–21.
12. National Research Council. Genetic vulnerability of major crops. Washington, D.C.: National Academy of Sciences, 1972.
13. Hughes JB, Daily GC, Ehrlich PR. Population diversity: its extent and extinction, Science 1997;278:689–92.
14. Black FL. Infectious diseases in primitive societies. Science 1975;187:515–8.
15. *Idem,* Why did they die? Science 1992;258:1739–40.

[16] Ormerod VTE. Ecological effect of control of African trypanosomiasis. Science 1976;191:815–21.
[17] Eisenberg L. The social construction of the human brain. Am J Psychiatry 1995;152:1563–75.
[18] Shatz CJ, Stryker MP. Prenatal tetrodotoxin infusion blocks segregation of retinogeniculate afferents. Science 1988;242:87–9.
[19] Penn AA, Riquelme PA, Feller MB, Shatz CJ. Competition in retinogeniculate patterning driven by spontaneous activity. Science 1998;279:2108–12.
[20] Catalano SM, Shatz CJ. Activity-dependent cortical target selection by thalamic axons. Science 1998;281:559–62.
[21] Wiesci TN. The postnatal development of the visual cortex and the influence of environment (the 1981 Nobel Prize lecture). Stockholm, Sweden: Nobel Foundation, 1982.
[22] Nelson CA, Bloom FE. Child development and neuroscience. Child Dev 1997;68:970–87
[23] Benes FM, Turtle M, Khan Y, Farol P. Myelination of a key relay zone in the hippocampal formation occurs in the human brain during childhood, adolescence, and adulthood. Arch Gen Psychiatry 1994;51:447–84.
[24] Eriksson PS, Perfileva E, Björk-Eriksson T, et al. Neurogenesis in the adult human hippocampus. Nat Med 1998;4:1313–7
[25] Wang X, Merzenich MM, Sameshima K, Jenkins WM. Remodeling of hand representation in adult cortex determined by timing of tactile stimulation. Nature 1995;378:71–5.
[26] Hamdy S, Bothwell JC, Aziz Q, Singh KD, Thompson DG. Long-term reorganization of human motor cortex driven by short-term sensory stimulation. Nat Neurosci 1998;1:64–8.
[27] Schlaug G, Janke L, Huang Y, Steinmetz H. In vivo evidence of structural brain asymmetry in musicians. Science 1995;267:699–701.
[28] Sterr A, Muller MM, Elbert T, Rockstroh B, Pantev C, Taub E. Perceptual correlates of changes in cortical representation of fingers in blind multifinger Braille readers. J Neurosci 1998;18:4417–23.
[29] Yang TT, Gallen CC, Ramachandran VS, Cobb S, Schwartz BL, Bloom FE. Noninvasive detection of cerebral plasticity in adult human somatosensory cortex. Neuroreport 1994;5:701–4.
[30] Mogilner A, Grossman JA, Ribary U, et al. Somatosensory cortical plasticity in adult humans revealed by magnetoencephalography. Proc Natl Acad Sci U S A 1993;90:3593–7.

Reflections on Dolly: What Can Animal Cloning Tell Us about the Human Cloning Debate?

Autumn Fiester

Introduction

The issue of human reproductive cloning has received a great deal of attention in public discourse. Bioethicists, policymakers, and the media have been quick to argue, almost unanimously, for an international ban on such attempts. Meanwhile, scientists in labs across the globe have proceeded with extensive research agendas and incredible progress in the cloning of animals. What in 1997 was considered a remarkable feat—the cloning Dolly the Sheep in Scotland—is today becoming almost commonplace. To date, scientists have successfully cloned many other species including a cat, a gaur, a rabbit, cows, mice, goats, pigs, mules and, most recently, a horse. In the works are projects to clone primates, dogs and a host of endangered species. Yet against this backdrop of a rapidly advancing science, there has been little attention paid to the bioethics of animal cloning, despite the ethical lessons that we might glean in the human case from reflections on that science.

An overview of the ethics of animal cloning sets out two different types of ethical consequence-based arguments and principle-based ones. To see how these anti-cloning arguments play out in the case of animal cloning, we must review the state of cloning science to see just how far we have come since Dolly made her debut.

The State of Animal Cloning Science

The progress in animal cloning science has been exponential. The first successful attempt at cloning was made by scientists at the Roslin Institute in Scotland just eight years ago in 1995, when they cloned two lambs, Megan and Morag. The lambs were cloned from cells from an early embryo. Two years later, the same scientists cloned Dolly the Sheep, this time beginning with the cell of an adult sheep. In the last six years, scientists have not only cloned other species, but they advanced the science of cloning to now include genetic modifications that serve particular pharmaceutical or agricultural purposes. For example, at the Roslin Institute, scientists have cloned transgenic sheep engineered to secrete a human protein that could be given to patients who lack it, such as hemophiliacs.[1] Along the same lines, transgenic goats are being cloned to produce a protein in their milk that researchers hope could be used to treat heart attacks and strokes.[2] One goal of cloning research is the production of genetically modified animal organs that are engineered to be compatible with a human recipient or the cloning of animals engineered to carry human genetic diseases so that they can make better models of the disease for research.[7] Agricultural researchers have similar projects underway: the cloning of animals with genes that make their milk or meat healthier for consumers, or the cloning of animals which are disease-resistant or reduce the environmental burden. Scientists are working on cloning goats with less fatty milk, chickens with no feathers to reduce the environmental costs of poultry farming, and pigs whose manure has less phosphorus and helps reduce environmental pollution.[4] At Texas A & M University, scientists have cloned cows resistant to brucellosis.[5]

Animal cloning may also soon play a role in American sports. One project at Texas A & M hopes to clone deer bucks with larger antlers, which will be attractive to hunters.[6] Once perfected, cloning techniques could conceivably be used to clone animals used in any competitive sport, like race horsing.

Not all cloning science is aimed solely at human ends. Many animal cloning projects are motivated by regard or concern for animals as ends in themselves. For example, there are many projects aimed at cloning endangered or even extinct species. A gaur was cloned but died in infancy,[7] and there are projects in progress to clone the extinct thylacine, Asian cheetah, and wooly mammoth.[8] One of the most famous cloning projects, the Missyplicity Project, was created to clone a beloved dog who had died.[9] Other cloning projects seek to create animals with fewer health problems, which could lead to healthier animal lives and greater animal welfare.

The Ethical Issues in Animal Cloning

Given the diverse motivations and types of animal cloning, the ethical terrain here is complex. In summary, animal cloning raises two types of moral problems: it may have negative consequences, to animals, human beings, or the environment; and, it may violate important moral prohibitions or principles.

The negative consequences to animals can be both narrowly and broadly construed. Narrowly construed, focusing on animals involved in cloning procedures, the most serious consequence is the pain and suffering they experience in the cloning process. More broadly construed, the negative consequences to animals include the deleterious effects of cloning on other populations of animals, such as livestock, unwanted pets, or endangered species. Human beings may be adversely affected by animal cloning either through the slippery slope of perfecting reproductive cloning techniques on animals and then applying them to human reproductive cloning, or by compromising the safety of the livestock used in food production. In both the areas of agricultural cloning and cloning for conservation, cloned animals may have a serious impact on the environment, either by breeding with non-clones or due to some unforeseen expression of a gene that has ramifications for the larger ecosystem.

Animal cloning might also be criticized on deontological grounds. Here there are ethical concerns about "playing God," the intrinsic value of the animals, and the objectification and commodification of animals. In the area of pet cloning, there is the potential for fraud or false-promising: grieving pet owners may be misled into believing that cloning will resurrect their beloved pet, and they may commit to storing their pet's DNA without understanding the true costs of cloning when the technique becomes commercially available. In the area of cloning sport animals, specifically race horses, critics have raised concerns about competitive fairness, arguing that cloning could ruin the sport.

Consequence-Based Arguments against Animal Cloning

Perhaps the most compelling argument against animal cloning is the very real suffering endured by animals involved in this science. There are four areas of concern with regard to the pain and suffering animals experience due to animal cloning: the suffering animals undergo during cloning procedures; the obstetrical complications that occur in the surrogate animal; the health of cloned animals; and the suffering animals will be forced to endure if cloned to exhibit, for research purposes, certain diseases and pathologies.

Recent data on the success rates of cloning procedures and the health and survival statistics of animal clones present a fairly grim picture. There is a large body of literature citing high rates of miscarriage, stillbirth, early death, genetic abnormalities, and chronic diseases among cloned animals. These problems occur against a backdrop of what in cloning science has been called "efficiency," the term used to talk about the percentage of live offspring from the number of transferred embryos. The efficiency of animal cloning has typically been about 1–2%, so for every 100 embryos that are implanted in surrogate animals, about 98% of the embryos fail to produce a live animal offspring.[10] Even when efficiency rates are at their best, the overwhelming majority of attempts fail. One study explicitly touting a "highly efficient" method for cloning pigs claims efficiency rates of only 5–12%.[11] This is still a failure rate of between 88–95%. These numbers have serious consequences for both the donor and surrogate being impregnated: surgery must be used to remove the donor animal's eggs and then another highly invasive technique used to implant the embryos into the surrogate. In the least "efficient" processes, for every two live cloned offspring, 100 eggs must be harvested and 100 embryos implanted. For unknown reasons, cloned fetuses often exhibit a very high birthweight, frequently necessitating a C-section delivery, again causing pain and suffering to the surrogate animal.

Of the live clones born, many experience compromised health status or early death. In one study of cloned pigs, researchers reported a 50% mortality rate for the live offspring, with 5 out of 10 dying between 3–130 days of age from ailments including chronic diarrhea, congestive heart failure, and decreased growth rate.[12] A study published last year in *Nature Genetics* showed that cloned mice experience early death due to liver failure and lung problems.[13] Another study showed that cloned mice had a high tendency to morbid obesity.[14]

These adverse effects on the animals involved in cloning procedures have prompted national animal welfare organizations to take a strong stance against animal cloning. The US Humane Society, for example, has requested a ban on products coming from cloned animals or their offspring.[15] Michael Appleby, the HSUS vice president for farm animals, argues that the animal welfare problems that already exist in factory farming will be exacerbated with the new biotechnology, for example, disease vulnerability. He argues, "Already animals are suffering from maladies at a rate unheard of before we applied biotechnology to the barnyard. It would be disastrously premature to put this technology into commercial practice."[16]

Proponents of animal cloning argue that efficiency rates are constantly improving and other studies show much better health outcomes for the cloned animals. One study on the health of cloned cattle reported normal growth rates of the cloned offspring and concluded that the surviving clones in this study were healthy two years after birth.[17] A closer look at the findings, however, shows that there were only 106 live births for 2170 implanted embryos, and of the 106 live offspring born, 24 of the calves died soon after birth; 11 of these 24 had severe physiological abnormalities, including digestive abnormalities, skeletal problems, deformities in the urinary track, or respiratory failure.[18] *Nature Biotechnology* recently published a metareview of the health status of clones from prior studies, and it reports that 77% of cloned animals showed no developmental abnormalities throughout the period of follow-up, though the percentage of healthy clones ranged across studies from 20–100%.[19] It is clear that the data varies widely across studies, but even in those with the highest efficiency rates and best health outcomes there is a significant cost in animal welfare of cloning science in its current stage of development.

That said, proponents of animal cloning argue that the net good of animal cloning science will far outweigh these costs to animal welfare. In a classic balancing of means and ends, proponents cite the potential benefits of animal cloning to human medicine, food production, pharmaceutical applications, and even to animal species themselves. While the pain and suffering animals endure from cloning procedures might be justified by noble ends (such as curing human or animal diseases, for example, or preserving endangered species), animal welfare issues are harder to ignore when cloning is undertaken for ends that seem frivolous or even, by some lights, inhumane. A prime example would be the quest to clone deer with larger antlers so that they are more attractive to hunters.[20]

Another rejoinder proponents make to their objectors is that cloners and cloning science should not be held to a higher standard of preserving animal welfare than is the current practice in the various areas of animal use. On this argument, we have already decided as a society that animals may be eaten, hunted, experimented on and confined. Whatever the standards are for humane treatment of animals in the different arenas in which they figure in modern life–agricultural, research, sport, etc.–those must be the same standards, and no higher, that govern the realm of animal cloning. Mark Greene calls this the "Accepted Practice Standard."[21] Animal cloning science ought not to be condemned for causing animals pain and suffering if that same level of pain and suffering is morally permissible in areas outside biotechnology research. By extension of this argument, we ought not to selectively condemn cloning researchers for cloning animals destined to live their whole lives in cages if we accept that other research or agricultural animals fare no better. Of course, opponents of animal cloning won't easily be caught in a contradiction here: they are very likely to be highly critical of our "standard" practices with regard to animal treatment if animal pain and suffering is what motivates their anti-cloning stance.

A second consequence-based argument against animal cloning involves the consequences to animals much more broadly construed, focusing on the effect of cloning on animals not involved in cloning procedures. Opponents of pet cloning, for example, cite the millions of unwanted pets in the United States as an argument against research intended to produce what are, in effect, identical twins of deceased pets. In support of this concern, the data on the number of companion animals euthanized in American shelters are shocking. The National Council on Pet Population Study and Policy found that in 1997 alone, 2,329,978 dogs and 1,759,743 cats entered shelters, and between 50–70% of these animals were euthanized.[22] According to the ASPCA, the numbers of animals entering shelters is much higher now. By their estimates, 8–12 millions companion animals enter shelters, and 60–70% are euthanized.[23] A similar number is cited in the 2001 Human Society report on the state of animals in the US. According to that report, 4–6 million dogs and cats were euthanized in shelters in 2001.[24] These figures do not include the millions of stray animals in the country: the ASPCA estimates that 70 million stray dogs and cats live in the US.[25] Opponents of pet cloning make two anti-cloning arguments: first, pet owners so devoted to their animal companions that they would spend thousands of dollars to produce their beloved pets' identical twin are precisely the type of adoptive parents who could save an already-existing animal's life through pet adoption, sparing one more dog or cat in the US from euthanasia; and, second, the money and energy now being spent on the development of pet-cloning, like the $3.7 million-dollar Missyplicity Project,[26] is a terrible waste of financial and intellectual resources that could much better serve animal welfare interests if redirected to better causes. Champions of pet cloning, including cloning scientists and investors, are quick to point out there will only ever be a small number of pets cloned even after the technology is perfected and widely available. On this argument, the number of animals who would be spared euthanasia in shelters if pet owners chose pet adoption instead of cloning would not have any measurable effect on the tragic numbers of animal-euthanasia.

In response, objectors could argue that this argument misses the mark: the criticism assumes that only one animal life is at stake either way, either one cloned animal or one non-cloned. But the opponent to pet cloning is making a utilitarian argument about distributive justice: the $20,000 or so that one will have to spend to create one non-clone could save the lives of literally thousands of non-clones. It also speaks to the moral status of animals: if animals have an intrinsic worth that merits this kind of expenditure, then it is criminal that we would allow so many of the same creatures to die; resources simply need to be redistributed in a way that makes better ethical sense. Of course, proponents of pet cloning will point out that the money in question can't in fact be re-distributed like this: pet owners who are willing to shell out $20,000 to clone their own beloved pet are not going to use that money to fund animal shelters should they decide not to go ahead with the cloning.

Two other consequence-based arguments can be made against animal cloning in general: it may have negative consequences for human beings, and it may have negative consequences for the environment. The most pressing concern about the way that animal cloning will affect human beings is through the slippery slope of reproductive cloning progress. Many express fear that once reproductive cloning techniques are perfected in primates, it will be a very short time until those techniques are applied to human beings. There are already reports that maverick scientists are experimenting with human reproductive cloning, and animal cloning techniques create a possible blueprint for accomplishing this terrifying feat.

A second concern about the effect of animal cloning on human beings involves food production from cloned livestock. The obvious question here is part ethics and part science: will this food be safe to eat? In late October 2003 the FDA issued a preliminary statement claiming that there was no evidence that food and milk from cloned animals was unsafe.[27] But a few days later, on November 4, 2003, they issued another statement claiming that they lacked enough information to state that it *was* safe.[28] The FDA predicts that it will be late 2004 before they can issue a recommendation about whether products from cloned animals need government approval before going to market.[29] Meanwhile, the current voluntary moratorium on releasing these products into the market remains in place. Many watchdog groups believe the FDA has underestimated consumer concerns about eating these products, and that wariness may not change even if the data demonstrate food safety. Consumers are already suspicious of GMO foods, especially in certain markets, and food producers will have a high bar to reach in assuring the public that agricultural products from cloned animals are not only safe but appealing. Most significantly in this debate is the fact that there really is very little research to date on the safety of these products, as the FDA admits—and there is certainly no long-term data on its possible effects.

Concerns about the environment involve two different categories of animal cloning, cloning for conservation purposes (the cloning of endangered or extinct species) and the cloning of livestock. Both concerns are motivated by the effect of clones on the ecosystem. The question here: will there be an adverse effect or an unanticipated impact on the environment from this interaction of clones with the environment? With the "recreation" of extinct species this question might be even more pressing than with endangered ones. It is one potential problem to have clones breed with non-clones, but another ecological problem altogether with recreating species that no longer exist.

Principle-Based Arguments against Animal Cloning

Apart from the ethical issues that arise because of the potential consequences of animal cloning, there is a significant set of ethical concerns about cloning that are based on deontological considerations. For this set of ethical issues, animal cloning appears to violate an important moral prohibition or duty. This is the type of argument some have called an "intrinsic" objection,[30] in which the activity is objectionable in and of itself, irrespective of the consequences.[31]

The most well-known of these deontological arguments is the "playing God" problem: ought we to be creating life in this matter? Is it our place to do this? On this view, cloning is a hubristic attempt by human beings to be divine. Cloning (and many types of genetic engineering[32]) crosses an important moral line between facilitating the creation of life (as in assisted reproduction) and engineering life. In the first case, human beings are the builders using a God-given blueprint; in the second, they are the architects. While many individuals making this type of argument come from a religious tradition, thus anchoring their arguments in theology, not all of them do. There is also a secular version of this moral concern. On this objection, we dehumanize ourselves and devalue the natural world by engaging in such activity. For these opponents of cloning, there is a reverence due the natural world that is violated when we manipulate living organisms in this manner. The sense that something is being lost or something profound is being compromised by cloning is part of what lends credence to Leon Kass's famous, but underdeveloped, "wisdom of repugnance" claim.[33] Individuals in this camp emphasize the widespread intuition that something is deeply wrong with this enterprise, however difficult it is to say precisely what the moral prohibition is. They point to cases like the Chicago-based artist, Eduardo Kac, who commissioned the creation of a transgenic rabbit that is green and glows under special lighting.[34] Kac's goal "is a new art form based on the use of genetic engineering to transform natural or synthetic genes to an organism, to create unique living beings."[35] Opponents of cloning argue that the irreverence shown by making living beings into art objects[36] shows us unfit to play the architect's role.

Proponents of animal cloning offer two rejoinders to these claims. First, they argue that we have been modifying organisms throughout human history. Modification of animals and, in the process the creation of new species, is not new.[37] What has changed is the means of modification and, possibly the scope. But if the modifications of the past were morally permissible, what is the morally relevant difference in the current science? If it is the means of the modification, then the real argument is about the negative consequences to the animals, not a deontological prohibition against playing God.

A second rejoinder made by cloning proponents is that the objectors miss the respect for and awe of nature that many cloning scientists feel. They argue that unlocking the mysteries of the natural world only makes them more humbled by those complexities. On their view, the scientific enterprise is sanctioned by God, evidenced by the intellectual acuity humankind was granted by their Creator that motivates scientific curiosity. Gary Comstock, for example, coming from the perspective of an evangelical Protestant, argues that God wants human beings to pursue science and he believes God endorses scientific endeavors like biotechnology.[38] It isn't playing God; it's doing what God has given us the mental gifts to do.

A second deontological argument is that cloning negates the intrinsic value of animals through both objectification and commodification. On this view, cloning treats animals as mere things, rather than living, breathing, sentient subjects. While admittedly animals are already considered property and products, cloning takes this objectification to new levels. Life for animals in agriculture, research, and the pharmaceutical industries is already bad enough, it's argued; cloning will desensitize us further from the suffering of these entities, placing them even more firmly in the "thing" category. If animals are simply objects that we can create, then how different are they from disposable products like automobiles, telephones, etc.? When animals were first being cloned in the mid to late '90s, the animals were given personal names, emphasizing their subjecthood and uniqueness. More and more, cloned animals are referred to by impersonal numbers and letters, rather than names. Writes Andrea Bonnicksen, "The fact that Polly, Morag, Megan,[39] and other creatures with personal monikers are fading in the wake of the impersonally named ACT calves[40] foretells a normalcy of genetic and cloning combinations in biotechnology."[41] With this normalcy comes acceptance of the objectification of these living beings. Of course, the proponents' argument is already made by the opponents: animals are now and have historically been considered objects, both legally and in practice, so cloning science simply reaffirms the status quo; it does not change the status of animals. Livestock and research animals are already referred to by numbers, not names. This is acceptable practice in non-cloning breeding and therefore constitutes no argument specific to the cloning of animals.

By extension of the objectification argument, opponents argue that cloning further commodifies animals. They are already seen as products for purchase, but, pre-cloning, they are—at least sometimes—seen as ends in themselves as well. Cloning connotes "pure product" and implies that no attention need to be paid to them as subjects. Opponents ask where this ultra-commodification will take us in terms of animal welfare. Again, the obvious response by proponents is that human beings have always bartered and traded animals as commodities, but, they add, at no time in human history has more concern been paid to the pain and suffering of food and research animals as is being paid to them today. That response may not be the end of the debate. It does not defeat the objection to argue that animals have always been a "means" for human beings. The distinction opponents may be trying to make is one between being a pure means and a means while also at the same time an end. If, for example, traditional farmers argued that they "cared" about their animals, then, even if they sold them or had them butchered for food, they could plausibly have been described as treating the animals as subjects or ends in themselves—while at the same time being a means for the farmer to sustain his life, physically or economically. This treatment of food animals as ends in themselves may have ended with the rise of factory farming, but does cloning exacerbate this trend? Does it delude us into believing that animals are no different from machines? Is there a reductionism of animal life that we have never heretofore seen?

On a completely different type of deontological argument, specific types of animal cloning are criticized on grounds of fraud or false promising. Pet cloning, in particular, is vulnerable to this type of objection. Opponents argue that grieving pet owners are deceived into believing that cloning is a way of resurrecting a deceased and beloved pet. This is a type of false advertising since cloning the animal would never replicate a lost pet due to the many environmental factors that influence personality and even appearance ("CC," the cloned calico cat, does not look like the mother). The business of pet cloning assumes genetic determinism, which is false. Pet cloning websites do not explicitly say that cloning will only produce an identical twin. On the Genetic Savings & Clone website, for example, they say, "Before gene banking your pet, we urge you to answer one question as honestly as possible: Do I want to bank my pet's DNA because I'm distraught and want the SAME pet back, or because my pet had a special genetic endowment that ought to be preserved? ...If your honest answer is that you are grieving your pet's loss and seeking an identical replacement, then we respectfully discourage you from using our services."[42] At rival Perpetuate's website, customers are explicitly told that the cloned animal will be an identical twin, although they cryptically cite the following reason to clone the animal: "The possibility of cloning a valued pet provides its owner with a degree of hope."[43] At Lazaron's site, they say: "Does your loved animal's singular genetic character have to die? No. IT CAN LIVE ON," and the web address is www.lazaron.cm/savinglife.html, implying that cloning is a way to save the original pet.[44]

A second issue of potentially misleading clients is that the cost of the actual cloning procedure is not easy to find or isn't on the websites of the three major pet-DNA banking companies, Genetic Savings and Clone, Lazaron, or Perpetuate. Under "Pricing" these companies discuss the price of the DNA banking, not the cloning of the pet, which has been estimated at $20,000 or more.[45] For example, at the Genetic Savings and Clone website this information is buried in the FAQ section. Instead of quoting a dollar amount in numbers, they say: "After Missy is cloned, our hope is to offer dog and cat cloning services in the low five figures, and to drop the price further as we increase our efficiency."[46] At the Perpetuate and Lazaron sites, consumers are not even given an estimate of the cost of the actual cloning.

Conclusion

Bioethicists, policymakers, and the public have too long neglected the important ethical issues raised by the science of animal cloning, with its implications for both animals and human beings alike. Given both the consequence-based and principle-based objections to animal cloning, it is evident that this science raises serious moral concerns and reservations that proponents of animal cloning must address. If animal cloning raises this level of ethical concern, how much more caution ought we to have with the science of human cloning? We have at present a wholly unregulated industry of animal cloning whose progress may have far-reaching and unanticipated consequences, well beyond the current projects under which the science is currently conducted. Animal cloning is a scientific enterprise that should no longer go under the public's radar.

Notes

[1] Scientists at the Roslin Institute have already successfully cloned transgenic sheep with this added protein. See E. Pennisi, "Transgenic Lambs From Cloning Lab," *Science* 277 (1997), p. 631. See also CNN News, "Report: Cloned Sheep Has Human Gene," July 24, 1997. Available at http://www.cnn.com/TECH/9707/24/polly/index.html.

[2] Information on these goats available at: http://www.gene.ch/genet/1999/May/msg00002.html and http://www.standford.edu/~eclipse9/sts129/cloning/timeline.html.

[3] I. Wilmut, "Cloning for Medicine," *Scientific American*, December, 1998, pp.58–63.

[4] T. Thomas, "Cloned and Genetically Engineered Animals," The Humane Society of the United States website, available at: http://www.hsus.org/ace/15401.

[5] K. Phillips, "Disease-Resistent Bull Cloned at Texas A & M," December 18, 2000. Available at: http://www.tamu.edu/aggiedaily/press/020214cc.html.

[6] "Deer Are Next Cloning Candidates," AgBiotechNet. Available at: http:www.agbiotechnet.com.

[7] A gaur, a type of wild ox on the verge of extinction, was

cloned in January, 2001, but the calf died soon after birth. Researchers say the bull died from dysentery unrelated to the cloning process. Information available at: http://www.bbc.co.uk/science/genes.

[8] "Cloning Hopes for Extinct Species" and "Project to Clone Extinct Cheetah Gets a Boost," AgBiotechNet. Available at: http:www.agbiotechnet.com.

[9] A dog was found to be too difficult to clone, so this is the project that eventually resulted in "CC" the calico cat.

[10] A. Coleman, "Somatic Cell Nuclear Transfer in Mammals: Progress and Application, *Cloning* 1 (1999), 185–200. See also, L. Paterson, "Somatic Cell Nuclear Transfer (Cloning) Efficiency," available at: http://www.roslin.ac.uk/public/webtablesGR.pdf.

[11] S. Walker, et al. "A Highly Efficient Method for Porcine Cloning By Nuclear Transfer Using *In Vitro*-Matured Oocytes," *Cloning and Stem Cells* 4 (2002), 105–112.

[12] A.B. Carter, et al. "Phentyping of Transgenic Cloned Pigs," *Cloning and Stem Cells*, 4 (2002), 131–145.

[13] N. Ogonuki, et al., "Early Death of Mice Cloned from Somatic Cells," *Nature Genetics* 30 (2002) 253–4.

[14] K. Tamashiro, "Cloned Mice Have an Obese Phenotype Not Transmitted to Their Offspring," *Nature Genetics* 8 (2002), 262–7.

[15] HSUS, "HSUS Asks the FDA to Ban Sales of Products from Cloned Farm Animals," October 9, 2002. Available at http://www.hsus.org/ace/15431.

[16] HSUS, "HSUS Asks the FDA to Ban Sales of Products from Cloned Farm Animals," October 9, 2002. Available at http://www.hsus.org/ace/15431.

[17] M. Pace, et al. "Ontogeny of Cloned Cattle to Lactation," *Biology of Reproduction*, 67 (2002), 334–9.

[18] M. Pace, et al. "Ontogeny of Cloned Cattle to Lactation," p.335.

[19] J.B. Cibelli, et al. "The Health Profile of Cloned Animals," *Nature Biotechnology* 20 (2002), 13–14.

[20] "Deer Are Next Cloning Candidates," AgBiotechNet. Available at: http:www.agbiotechnet.com.

[21] M. Greene, "New Dog: Old Tricks," *Journal of Applied Animal Welfare*, 3 (2002), 239–242.

[22] National Council on Pet Population Study and Policy, "Shelters Statistics Survey," 1994–7. Avail at: http://

www.petpopulation.org/statsurvey.html.

[23] ASPCA, "Annual Shelter Statistics." Available at http://www.aspca.org.

[24] P.G. Irwin, "Overview: The State of Animals in 2001," in *The State of Animals 2001*, eds. DJ Salem and AN Rowan, (Washington DC: Humane Society Press, 2001).

[25] ASPCA, "Annual Shelter Statistics." Available at http://www.aspca.org.

[26] The Missyplicity project was undertaken to attempt to clone a mixed-breed dog, "Missy," who had died. This is the project that eventually created "CC" the calico cat, when cloning a dog proved too difficult. Http://www.tamu.edu/aggiedaily/press/020214cc.html

[27] L. Neergaard, "FDA: Cloned Animal Meat Appears Safe," *Washington Post*, October 31, 2003. Available at: http://washingtonpost.com/ac2/wp-dyn/A46933-2003Oct31?language=printer.

[28] Associated Press, "FDA Wants More Data on Cloned-meat Safety," *Washington Post*, November 4, 2003. Available at: http://washingtonpost.com/ac2/wp-dyn/A867-2003Nov4?language=printer.

[29] "FDA Wants More Data on Cloned-meat Safety," *Washington Post*, November 4, 2003.

[30] See, for example, G. Comstock, "Ethics and Genetically Modified Foods," Testimony for the New Zealand Royal Commission on Genetic Modification, September 10, 2001, p. 6. Available at http://www.biotech.iastate.edu/publications/IFAFS/NewZealand_paper.pdf.

[31] Although it makes for a neat and clean distinction, deontological concerns are not consequence-neutral. All serious deontologists are concerned about the real-world effects of moral actions, which, on their view, is what explains how we get the prohibition in the first place.

[32] Jeremy Rifkin, for example, made this argument in 1983 in reference to the science of genetically modifying animals, what he calls "algeny." See J. Rifkin, *Algeny* (New York: Viking Press, 1983).

[33] L. Kass, "The Wisdom of Repugnance: Why We Should Ban the Cloning of Humans," *infra*.

[34] U. Allmendinger, "One Small Hop for Alba, One Large Hop for Mankind," *NY Arts Magazine* 6 (2001); and, N. Boyce, "Pets of the Future," *U.S. News & World Report*, March 11,

2002.

[35] E. Kac, "GFP Bunny." Available at: http://www.ekac.org/gfpbunny.html#gfpbunnyanchor.

[36] Kac would disagree with this depiction of the project, protesting, "I have never thought of Alba [the transgenic rabbit] as an art object in the sense that one would create a sculpture or a painting. It's not about making an object. I invent situations." See, Boyce, *U.S. News & World Report*, p.48. Critics respond, "*Res ipsa loquitur.*"

[37] See, for example, J. Burkhardt, "The Inevitability of Animal Biotechnology? Ethics and the Scientific Attitude," in *Animal Biotechnology and Ethics*, eds. A. Holland and A. Johnson (London: Chapman & Hall, 1998), pp. 114–131.

[38] G. Comstock, "Ethics and Genetically Modified Foods," p.6.

[39] Two lambs, Megan and Morag, were the first successful cloning attempts of the Roslin Institute in Edinburgh Scotland in the summer of 1995. They were cloned from cultured cells from a nine-day-old embryo. In contrast, Dolly the sheep was cloned in 1996 from an adult ewe's cell. Polly, born in the summer of 1997, was the first transgenic sheep, also created at the Roslin Institute. Polly was genetically engineered with a human gene to secrete a human protein in her milk that could be used to treat hemophilia.

[40] ACT3, ACT4, and ACT5 are the "names" of three calves cloned in 1998 in an experiment designed to increase the efficiency of the cloning process and make it feasible to do large-scale cattle cloning. J. Cibelli, S., et al. "Cloned Transgenic Calves Produced from Nonquiescent Fetal Fibroblasts, " *Science* 280 (1998), p.1256.

[41] A Bonnicksen, "First Dolly, Now Polly: Policy Implications of the Birth of a Transgenic Cloned Lamb," in *The Cloning Sourcebook*, ed. A. Klotzko (New York: Oxford University Press, 2001), p.267.

[42] "Is Cloning Right for You?" Genetic Savings and Clone, Inc. Available at: http://www/savingsandclone.com/services/right_for_you.php.

[43] http://www.perpetuate.net/index.htm.

[44] http://www.lazaron.cm/savinglife.htm.

[45] "The Making of Fido 2.0" MSNBC. Available at: http://www.msnbc.com/site_elements/blank.html.

[46] FAQ, "When Pet Cloning Is Possible, How Much Will It Cost?", Genetic Savings and Clone, Inc. Available at: http://www/savingsandclone.com/faqs/general_faq.php.

The Uncertain Art: Narcissus Looks into the Laboratory

Sherwin Nuland

Among the many ancient Greek medical writings that scholars of an earlier generation credited to Hippocrates, there is a particularly interesting essay that deals with the education of a physician. "Law"'s brief length of only five paragraphs, and certain of its trenchant observations, caused one translator to call it a "quaint little piece" and voice regret that "it has strangely been neglected by scholars." His concern was no doubt based on the fact that this particular gem contains more good sense per line than any of the approximately sixty treatises once thought to have been written by the Father of Medicine himself.

Some of the statements in "Law" are more aphoristic than the vast majority of those in *The Aphorisms* and exceed them in universality. And as for being timeless—well, try this one, about time itself: "It is time which imparts strength to all things and brings them to maturity." I found myself thinking about just this statement only a few days ago, while reading a newspaper account of yet another triumph in the world of molecular biology, this one being page-one-headlined by the exuberant *New York Times*: "In Big Advance in Cloning, Biologists Create 50 Mice."

Imagine that—the biologist as creator! Or, perhaps before too long, as Creator.

Not only did the researchers "churn out clones of adult mice," announced the *Times,* but they even went so far as to make clones of clones. One authority at Princeton University (to whom the article refers as a "mouse geneticist") "described the speed at which the cloning had progressed as breathtaking." He went further: " 'Absolutely,' he said, 'we're going to have cloning of humans.' " According to the mouse expert, even the safeguards of strict scientific protocol and the delay required to perfect the technique in monkeys would not prevent in vitro fertilization clinics from being able to add human cloning to their bag of tricks within five to ten years.

It's not enough that we are plunging pell-mell toward cloning one another. Even the genetic enhancement of laboratory-crafted people is now being talked about. This would mean quite a bit more than the current therapeutic aims of introducing genes into patients to fight or ward off disease or of cloning for the purposes of tissue and organ transplantation. Changes are now being considered that would improve the very germplasm, the permanent heredity, of these "created" clones. Traits thus made inherent would be potentially transferable to every succeeding generation. This goes beyond fantasizing about Bionic Man to conjuring up the dream of Designer Man.

The velocity of our head-over-heels rush to clinical fulfillment comes into perspective when we recall that only a bit more than a year and a half has passed since the newborn Dolly first made those appealing sheep's eyes at the television cameras. At that time, no serious scientist believed that an attempt at reproducing the feat in humans was possible within a decade. And now we are being told that a decade is the outer limit within which the process may be not only accomplished but even made available to the public just for the asking (and the paying).

Meanwhile, other researchers are less concerned with crafting an improved version of the present generation for transmission into the next than with playing around with a project that has been dear to mankind's heart ever since our species first made its appearance in dim prehistory. For these scientists, it is not sufficient to clone our bettered selves into interminable generations; they are experimenting with techniques that might have the potential to make some members of our generation themselves interminable. Their goal is the lengthening of life beyond any span that clinical and public health advances (which have already added some thirty-five years to our life expectancy during this century) might anticipate. What they are talking about now is an increase not only in expectancy, but in the very life span granted by nature and evolution to our species and every other.

Noting that a structure found at the end of the DNA molecule, called a telomere, decreases in size each time a cell divides, these researchers have been working with a gene that codes for the enzyme telomerase. Telomerase has the ability to maintain or even increase the length of the telomere. In the laboratory, such manipulation has resulted in a marked rise in the number of times a cell can divide before dying. As recently as two years ago, responsible molecular biologists scoffed at the idea that this work could extend the normal life span, but no longer. Telomerase is now a hot research topic.

There is something just a bit scary about the way researchers describe what they see in their experiments. They say the cells are rejuvenated, a word reminiscent of the age-old searchings for a source of renewed vigor that culminated earlier in this century in the "magical" effects of monkey gland extracts and other such nostrums for perpetuating youth.

Not only would some of the current rush to fruition have been deemed absurdly futuristic only a few years ago, but the often overheated media have flavored their reports of it with the promise of imminent wondrous applications to human happiness. Earlier this year, the selfsame *New York Times* reported telomerase's extension of cellular reproductive lines with a detailed story, in the lead paragraph of which was posed a tantalizing question: "If cells can be made to live indefinitely, can people be made immortal?" The large headline over the article read, "Longevity's New Lease on Life." Can anyone be blamed for believing that this fantasy is close to becoming reality?

The media are not without reason for enthusiasm. Daring statements are being made by otherwise cautious scientists about the implications of their work. As one after another of nature's hitherto closely guarded secrets is revealed by their relentless ingenuity, researchers have begun to allow themselves to think the previously unthinkable.

Some of their rumination, like that of the mouse geneticist, is public. As in his case, it generally takes the form of pithy quotes delivered over telephone lines in response to questing science writers, who then rush them into print for immediate consumption by a public eager to believe that a New Jerusalem of health and longevity is at hand. Perhaps egged on by a recently acquired prophetic image, many biomedical scientists have abandoned the restraint that has long characterized their breed. To the usual unbounded zeal for the next research step has been added a less usual unbounded zeal for the immediacy of clinical use. Because of a few awe-inspiring discoveries, many ordinary citizens have changed their opinion about how far we dare to look ahead and how fast we dare to go. Caught up in the infectious excitement of biomedical science and its commentators, we seem to have forgotten about the leavening effect of time's passage, and the maturity it can bring.

Our society has become very much like an overstimulated child. Perhaps such an analogy can be taken even further. The byproducts of biomedicine's brilliance have rubbed off on all of us, even those without the training or background to grasp fully the factual basis of the advances. The kind of child our society resembles just now is one whose intelligence far exceeds his maturity. Every teacher and every parent knows what a formula for disaster that can be. Among some of the scientists themselves, the brilliance-to-maturity ratio may be strikingly higher than among the general population, and not only because they tend to be smarter than the rest of us.

This might be the time to do just what wise teachers do when faced with such a situation in the classroom: a child whose intellectual attainments far outstrip his ability to deal with the consequences of being so smart, a child who is likely, therefore, to make mistakes in using his genius that will ultimately harm himself and others. What a wise teacher does with such a child is not to promote him to the next grade when the school year ends. The child stays where he is until his social abilities catch up to his brainpower. Years ago, it was the custom to push such brilliant kids forward rapidly through the grades, but the price—in psychological illness and even breakdowns—proved to be very high. Too many of these bright youngsters never fulfilled the promise of their intellectual gifts because they were not given the time to mature. Had they been kept back, they would have been allowed to grow up enough to comprehend what might happen as a result of their genius.

What I am suggesting here is a brake on the application of technologies whose consequences we can at present only begin to contemplate. For the first time in the annals of scientific research, we are faced with discoveries whose implications to society go far beyond the community of researchers, physicians, and those patients who would be directly involved. We may have reached the point where we can no longer afford to permit the scientists alone to determine how far to carry the applications of their work.

Some might even suggest that the *direction* of research, too, should be influenced by open discussion in our society. Though most of us would be reluctant to take such a position, the borderline between the basic science laboratory and the laboratory for product development has become so obscured that even such a radical proposition might one day need to be entertained.

But regardless of the stance taken in such matters, one obvious truth must be acknowledged: when an entire society is to be affected, an entire informed society should participate in the decisions. The society should slow down and give itself time to think, time to confer, time to mature.

The federal government has shown signs of understanding the present dilemma. In 1995, President Clinton formed the National Bioethics Advisory Commission to deal with the kinds of issues raised by the new branches of research. In what may prove to be its most significant decision thus far, this panel—composed of representatives from the fields of medicine, nursing, religion, law, ethics, industry, education, and health advocacy—did in effect decide to keep the child back a grade by supporting the president's planned five-year moratorium on the use of federal funds to attempt human cloning. In its ninety days of deliberation, during which the commission heard testimony from a cross section of informed societal interests, it came down on the side of banning application, while supporting the need for continuing research. Implicit in its recommendations are two guiding and interrelated notions: first, although we have knowledge aplenty, there would be no benefit—even if it were practically possible—in ceasing to pursue even more knowledge; second, knowledge without wisdom is a clear and present danger. The members of the commission did not have to add the lesson of "Law," that wisdom comes only with maturity and maturity comes only with time.

Law, if not "Law," may in fact soon enter the arena. Senators Feinstein of California and Kennedy of Massachusetts—

from sea to shining sea—have introduced a bill, S.R. 1611, that for ten years would prohibit attempts to clone a human being, while not restricting other kinds of cloning research, such as that on cells and tissues. The term of the proposed legislation is perhaps longer than even some moderate observers might wish, but events may permit it to be shortened. On the other hand, the term may need an extension once the manifold issues raised by the new science come more clearly into view.

If for no other reason, our society should at least pause briefly to think about its motivations for plunging forward without due consideration of the range of possible consequences or of the immediate financial cost of the research in a time of limited resources. The great advances in the real health of humanity in the past hundred years have occurred more as the result of such preventive measures as immunization, personal cleanliness, water purification, and other public health efforts than from any application—except perhaps antibiotics—of the technologies of cure to individual patients. These preventive measures—and not surgery, pharmacology, molecular biology, or the therapies given in hospitals—are the primary reason for our vastly improved life expectancy. It is health, after all, that should concern us. And it should concern us far more than the mere desire to perpetuate ourselves.

Prevention and early diagnosis are more direct routes to health and a good quality of life than any stratagem yet devised, or likely to be devised, in our lifetimes. As long as resources are finite, as they will always be, we would do well not to abandon the wisdom that has stood us in such good stead in the past. The magnificences served up daily by the molecular biologists may exhilarate us, but our real goal should not be the satisfaction of the egocentricity within each of us. To this end, we need to review not only our headlong dash into the arms of the researchers but our priorities as well.

The search for wisdom is always fraught with dangers, but those inherent in the present situation arise from the very fabric of human character. After all, who among us has not occasionally, or more than occasionally, cherished the notion that we might live forever, or that we might stay young far beyond our years, or that we might even continue to exist in some descendant who is, at least genetically, our duplicate? For some, these are more than just notions—they are a prevailing philosophy of life, even an obsession. Maybe this is why, when practical applications are only the glib predictions of overzealous researchers, keen votaries have uncritically hailed the future of genetic studies.

The prevailing mood of our time is self-absorption, and its natural extension, narcissism. Manifestations are everywhere apparent: the youth culture; the huge market in cosmetic surgery; the entire cult of "personal fulfillment"; the so-called human potential movement; the outsize emphasis on the individual as opposed to the cultural group and even the family; the popularity of new drugs to grow new hair or build new erections; and, finally, the very situation I am addressing here—the possibility that research in genetics will put us on the road toward eternal life. In pursuing vanity, far too many of us have simply lost hold of our senses. There has never been a period in human history during which the creed of self-indulgence has found so many apostles.

If there exists a single characteristic that unifies all self-absorption and vanity, it surely must be foolishness. Knowing lots of things does not make us any less foolish. Mere information is only the beginning of knowledge, and even knowledge does not of itself lead to maturity, nor does it guarantee good judgment. We need to grow up a lot more as a society before we are ready to play with the new toys being so efficiently made for us by the precocious scientists. Growing up takes time and reflection. T. S. Eliot reminds us that we have paid a high price for the compulsion to accelerate the engines of our existence:

> Where is the Life we have lost in living?
> Where is the wisdom we have lost in knowledge?
> Where is the knowledge we have lost in information?

The Wisdom of Repugnance: Why We Should Ban the Cloning of Humans

Leon Kass

Our habit of delighting in news of scientific and technological breakthroughs has been sorely challenged by the birth announcement of a sheep named Dolly. Though Dolly shares with previous sheep the "softest clothing, woolly, bright," William Blake's question, "Little Lamb, who made thee?" has for her a radically different answer: Dolly was, quite literally, made. She is the work not of nature or nature's God but of man, a British man, Ian Wilmut, and his fellow scientists. What's more, Dolly came into being not only asexually—ironically, just like "He [who] calls Himself a Lamb"—but also as the genetically identical copy (and the perfect incarnation of the form or blueprint) of a mature ewe, of whom she is a clone. This long-awaited yet not quite expected success in cloning a mammal raised immediately the prospect—and the specter—of cloning human beings: "I a child and Thou a lamb," despite our differences, have always been equal candidates for creative making, only now, by means of cloning, we may both spring from the hand of man playing at being God.

After an initial flurry of expert comment and public consternation, with opinion polls showing overwhelming opposition to cloning human beings, President Clinton ordered a ban on all federal support for human cloning research (even though none was being supported) and charged the National Bioethics Advisory Commission to report in ninety days on the ethics of human cloning research. The commission (an eighteen-member panel, evenly balanced between scientists and nonscientists, appointed by the president and reporting to the National Science and Technology Council) invited testimony from scientists, religious thinkers and bioethicists, as well as from the general public. It is now deliberating about what it should recommend, both as a matter of ethics and as a matter of public policy.

Congress is awaiting the commission's report, and is poised to act. Bills to prohibit the use of federal funds for human cloning research have been introduced in the House of Representatives and the Senate; and another bill, in the House, would make it illegal "for any person to use a human somatic cell for the process of producing a human clone." A fateful decision is at hand. To clone or not to clone a human being is no longer an academic question.

Taking Cloning Seriously, Then and Now

Cloning first came to public attention roughly thirty years ago, following the successful asexual production, in England, of a clutch of tadpole clones by the technique of nuclear transplantation. The individual largely responsible for bringing the prospect and promise of human cloning to public notice was Joshua Lederberg, a Nobel Laureate geneticist and a man of large vision. In 1966, Lederberg wrote a remarkable article in The

American Naturalist detailing the eugenic advantages of human cloning and other forms of genetic engineering, and the following year he devoted a column in The Washington

Post, where he wrote regularly on science and society, to the prospect of human cloning. He suggested that cloning could help us overcome the unpredictable variety that still rules human reproduction, and allow us to benefit from perpetuating superior genetic endowments. These writings sparked a small public debate in which I became a participant. At the time a young researcher in molecular biology at the National Institutes of Health (NIH), I wrote a reply to the Post, arguing against Lederberg's amoral treatment of this morally weighty subject and insisting on the urgency of confronting a series of questions and objections, culminating in the suggestion that "the programmed reproduction of man will, in fact, dehumanize him."

Much has happened in the intervening years. It has become harder, not easier, to discern the true meaning of human cloning. We have in some sense been softened up to the idea—through movies, cartoons, jokes and intermittent commentary in the mass media, some serious, most lighthearted. We have become accustomed to new practices in human reproduction: not just in vitro fertilization, but also embryo manipulation, embryo donation and surrogate pregnancy. Animal biotechnology has yielded transgenic animals and a burgeoning science of genetic engineering, easily and soon to be transferable to humans.

Even more important, changes in the broader culture make it now vastly more difficult to express a common and respectful understanding of sexuality, procreation, nascent life, family, and the meaning of motherhood, fatherhood and the links between the generations. Twenty-five years ago, abortion was still largely illegal and thought to be immoral, the sexual revolution (made possible by the extramarital use of the pill) was still in its infancy, and few had yet heard about the reproductive rights of single women, homosexual men and lesbians. (Never mind shameless memoirs about one's own incest!) Then one could argue, without embarrassment, that the new technologies of human reproduction—babies without sex—and their confounding of normal kin relations—who's the mother: the egg donor, the surrogate who carries and delivers, or the one who rears?—would "undermine the justification and support that biological parenthood gives to the monogamous marriage." Today, defenders of stable, monogamous marriage risk charges of giving offense to those adults who are living in "new family forms" or to those children who, even without the benefit of assisted reproduction, have acquired either three or four parents or one or none at all. Today, one must even apologize for voicing opinions that twenty-five years ago were nearly universally regarded as the core of our culture's wisdom on these matters. In a world whose once-given natural boundaries are blurred by technological change and whose moral boundaries are seemingly up for grabs, it is much more difficult to make persuasive the still compelling case against cloning human beings. As Raskolnikov put it, "man gets used to everything—the beast!"

Indeed, perhaps the most depressing feature of the discussions that immediately followed the news about Dolly was their ironical tone, their genial cynicism, their moral fatigue: "an udder way of making lambs" (*Nature*), "who will cash in on breakthrough in cloning?" (*The Wall Street Journal*), "is cloning baaaaaaaad?" (*The Chicago Tribune*). Gone from the scene are the wise and courageous voices of Theodosius Dobzhansky (genetics), Hans Jonas (philosophy) and Paul Ramsey (theology) who, only twenty-five years ago, all made powerful moral arguments against ever cloning a human being. We are now too sophisticated for such argumentation; we wouldn't be caught in public with a strong moral stance, never mind an absolutist one. We are all, or almost all, post-modernists now.

Cloning turns out to be the perfect embodiment of the ruling opinions of our new age. Thanks to the sexual revolution, we are able to deny in practice, and increasingly in thought, the inherent procreative teleology of sexuality itself. But, if sex has no intrinsic connection to generating babies, babies need have no necessary connection to sex. Thanks to feminism and the gay rights movement, we are increasingly encouraged to treat the natural heterosexual difference and its preeminence as a matter of "cultural construction." But if male and female are not normatively complementary and generatively significant, babies need not come from male and female complementarity. Thanks to the prominence and the acceptability of divorce and out-of-wedlock births, stable, monogamous marriage as the ideal home for procreation is no longer the agreed-upon cultural norm. For this new dispensation, the clone is the ideal emblem: the ultimate "single-parent child."

Thanks to our belief that all children should be wanted children (the more high-minded principle we use to justify contraception and abortion), sooner or later only those children who fulfill our wants will be fully acceptable. Through cloning, we can work our wants and wills on the very identity of our children, exercising control as never before. Thanks to modern notions of individualism and the rate of cultural change, we see ourselves not as linked to ancestors and defined by traditions, but as projects for our own self-creation, not only as self-made men but also man-made selves; and self-cloning is simply an extension of such rootless and narcissistic self-re-creation.

Unwilling to acknowledge our debt to the past and unwilling to embrace the uncertainties and the limitations of the future, we have a false relation to both: cloning personifies our desire fully to control the future, while being subject to no controls ourselves. Enchanted and enslaved by the glamour of technology, we have lost our awe and wonder before the deep mysteries of nature and of life. We cheerfully take our own beginnings in our hands and, like the last man, we blink.

Part of the blame for our complacency lies, sadly, with the field of bioethics itself, and its claim to expertise in these moral matters. Bioethics was founded by people who understood that the new biology touched and threatened the deepest matters of our humanity: bodily integrity, identity and individuality, lineage and kinship, freedom and self-command, eros and aspiration, and the relations and strivings of body and soul. With its capture by analytic philosophy, however, and its inevitable routinization and professionalization, the field has by and large come to content itself with analyzing moral arguments, reacting to new technological developments and taking on emerging issues of public policy, all performed with a naive faith that the evils we fear can all be avoided by compassion, regulation and a respect for autonomy. Bioethics has made some major contributions in the protection of human subjects and in other areas where personal freedom is threatened; but its practitioners, with few exceptions, have turned the big human questions into pretty thin gruel.

One reason for this is that the piecemeal formation of public policy tends to grind down large questions of morals into small questions of procedure. Many of the country's leading bioethicists have served on national commissions or state task forces and advisory boards, where, understandably, they have found utilitarianism to be the only ethical vocabulary acceptable to all participants in discussing issues of law, regulation and public policy. As many of these commissions have been either officially under the aegis of NIH or the Health and Human Services Department, or otherwise dominated by powerful voices for scientific progress, the ethicists have for the most part been content, after some "values clarification" and wringing of hands, to pronounce their blessings upon the inevitable. Indeed, it is the bioethicists, not the scientists, who are now the most articulate defenders of human cloning: the two witnesses testifying before the National Bioethics Advisory Commission in favor of cloning human beings were bioethicists, eager to rebut what they regard as the irrational concerns of those of us in opposition. One wonders whether this commission, constituted like the previous commissions, can tear itself sufficiently free from the accommodationist pattern of rubber-stamping all technical innovation, in the mistaken belief that all other goods must bow down before the gods of better health and scientific advance.

If it is to do so, the commission must first persuade itself, as we all should persuade ourselves, not to be complacent about what is at issue here. Human cloning, though it is in some respects continuous with previous reproductive technologies, also represents something radically new, in itself and in its easily foreseeable consequences. The stakes are very high indeed. I exaggerate, but in the direction of the truth, when I insist that we are faced with having to decide nothing less than whether human procreation is going to remain human, whether children are going to be made rather than begotten, whether it is a good thing, humanly speaking, to say yes in principle to the road which leads (at best) to the dehumanized rationality of *Brave New World*. This is not business as usual, to be fretted about for a while but finally to be given our seal of approval. We must rise to the occasion and make our judgments as if the future of our humanity hangs in the balance. For so it does.

The State of the Art

If we should not underestimate the significance of human cloning, neither should we exaggerate its imminence or misunderstand just what is involved. The procedure is conceptually simple. The nucleus of a mature but unfertilized egg is removed and replaced with a nucleus obtained from a specialized cell of an adult (or fetal) organism (in Dolly's case, the donor nucleus came from mammary gland epithelium). Since almost all the hereditary material of a cell is contained within its nucleus, the renucleated egg and the individual into which this egg develops are genetically identical to the organism that was the source of the transferred nucleus. An unlimited number of genetically identical individuals—clones—could be produced by nuclear transfer. In principle, any person, male or female, newborn or adult, could be cloned, and in any quantity. With laboratory cultivation and storage of tissues, cells outliving their sources make it possible even to clone the dead.

The technical stumbling block, overcome by Wilmut and his colleagues, was to find a means of reprogramming the state of the DNA in the donor cells, reversing its differentiated expression and restoring its full totipotency, so that it could again direct the entire process of producing a mature organism. Now that this problem has been solved, we should expect a rush to develop cloning for other animals, especially livestock, in order to propagate in perpetuity the champion meat or milk producers. Though exactly how soon someone will succeed in cloning a human being is anybody's guess, Wilmut's technique, almost certainly applicable to humans, makes attempting the feat an imminent possibility.

Yet some cautions are in order and some possible misconceptions need correcting. For a start, cloning is not xeroxing. As has been reassuringly reiterated, the clone of Mel Gibson, though his genetic double, would enter the world hairless, toothless and peeing in his diapers, just like any other human infant. Moreover, the success rate, at least at first, will probably not be very high: the British transferred 277 adult nuclei into enucleated sheep eggs, and implanted twenty-nine clonal embryos, but they achieved the birth of only one live lamb clone. For this reason, among others, it is unlikely that, at least for now, the practice would be very popular, and there is no immediate worry of mass-scale production of multicopies. The need of repeated surgery to obtain eggs and, more crucially, of numerous borrowed wombs for implantation will surely limit use, as will the expense; besides, almost everyone who is able will doubtless prefer nature's sexier way of conceiving.

Still, for the tens of thousands of people already sustaining over 200 assisted-reproduction clinics in the United States and already availing themselves of in vitro fertilization, intracytoplasmic sperm injection and other techniques of assisted reproduction, cloning would be an option with virtually no added fuss (especially when the success rate improves). Should commercial interests develop in "nucleus-banking," as they have in sperm-banking; should famous athletes or other celebrities decide to market their DNA the way they now market their autographs and just about everything else; should techniques of embryo and germline genetic testing and manipulation arrive as anticipated, increasing the use of laboratory assistance in order to obtain "better" babies—should all this come to pass, then cloning, if it is permitted, could become more than a marginal practice simply on the basis of free reproductive choice, even without any social encouragement to upgrade the gene pool or to replicate superior types. Moreover, if laboratory research on human cloning proceeds, even without any intention to produce cloned humans, the existence of cloned human embryos in the laboratory, created to begin with only for research purposes, would surely pave the way for later baby-making implantations.

In anticipation of human cloning, apologists and proponents have already made clear possible uses of the perfected technology, ranging from the sentimental and compassionate to the grandiose. They include: providing a child for an infertile couple; "replacing" a beloved spouse or child who is dying or has died; avoiding the risk of genetic disease; permitting reproduction for homosexual men and lesbians who want nothing sexual to do with the opposite sex; securing a genetically identical source of organs or tissues perfectly suitable for transplantation; getting a child with a genotype of one's own choosing, not excluding oneself; replicating individuals of great genius, talent or beauty—having a child who really could "be like Mike"; and creating large sets of genetically identical humans suitable for research on, for instance, the question of nature versus nurture, or for special missions in peace and war (not excluding espionage), in which using identical humans would be an advantage. Most people who envision the cloning of human beings, of course, want none of these scenarios. That they cannot say why is not surprising. What is surprising, and welcome, is that, in our cynical age, they are saying anything at all.

The Wisdom of Repugnance

"Offensive." "Grotesque." "Revolting." "Repugnant." "Repulsive." These are the words most commonly heard regarding the prospect of human cloning. Such reactions come both from the man or woman in the street and from the intellectuals, from believers and atheists, from humanists and scientists. Even Dolly's creator has said he "would find it offensive" to clone a human being.

People are repelled by many aspects of human cloning. They recoil from the prospect of mass production of human beings, with large clones of look-alikes, compromised in their individuality; the idea of father–son or mother–daughter twins; the bizarre prospects of a woman giving birth to and rearing a genetic copy of herself, her spouse or even her deceased father or mother; the grotesqueness of conceiving a child as an exact replacement for another who has died; the utilitarian creation of embryonic genetic duplicates of oneself, to be frozen away or created when necessary, in case of need for homologous tissues or organs for transplantation; the narcissism of those who would clone themselves and the arrogance of others who think they know who deserves to be cloned or which genotype any child-to-be should be thrilled to receive; the Frankensteinian hubris to create human life and increasingly to control its destiny; man playing God. Almost no one finds any of the suggested reasons for human cloning compelling; almost everyone anticipates its possible misuses and abuses. Moreover, many people feel oppressed by the sense that there is probably nothing we can do to prevent it from happening. This makes the prospect all the more revolting.

Revulsion is not an argument; and some of yesterday's repugnances are today calmly accepted—though, one must add, not always for the better. In crucial cases, however, repugnance is the emotional expression of deep wisdom, beyond reason's power fully to articulate it. Can anyone really give an argument fully adequate to the horror which is father-daughter incest (even with consent), or having sex with animals, or mutilating a corpse, or eating human flesh, or even just (just!) raping or murdering another human being? Would anybody's failure to give full rational justification for his or her revulsion at these practices make that revulsion ethically suspect? Not at all. On the contrary, we are suspicious of those who think that they can rationalize away our horror, say, by trying to explain the enormity of incest with arguments only about the genetic risks of inbreeding.

The repugnance at human cloning belongs in this category. We are repelled by the prospect of cloning human beings not because of the strangeness or novelty of the undertaking, but because we intuit and feel, immediately and without argument, the violation of things that we rightfully hold dear. Repugnance, here as elsewhere, revolts against the excesses of human willfulness, warning us not to transgress what is unspeakably profound. Indeed, in this age in which everything is held to be permissible so long as it is freely done, in which our given human nature no longer commands respect, in which our bodies are regarded as mere instruments of our autonomous rational wills, repugnance may be the only voice left that speaks up to defend the central core of our humanity. Shallow are the souls that have forgotten how to shudder.

The goods protected by repugnance are generally overlooked by our customary ways of approaching all new biomedical technologies. The way we evaluate cloning ethically will in fact be shaped by how we characterize it descriptively, by the context into which we place it, and by the perspective from which we view it. The first task for ethics is proper description. And here is where our failure begins.

Typically, cloning is discussed in one or more of three familiar contexts, which one might call the technological, the liberal and the meliorist. Under the first, cloning will be seen as an extension of existing techniques for assisting reproduction and determining the genetic makeup of children. Like them, cloning is to be regarded as a neutral technique, with no inherent meaning or goodness, but subject to multiple uses, some good, some bad. The morality of cloning thus depends absolutely on the goodness or badness of the motives and intentions of the cloners: as one bioethicist defender of cloning puts it, "the ethics must be judged [only] by the way the parents nurture and rear their resulting child and whether they bestow the same love and affection on a child brought into existence by a technique of assisted reproduction as they would on a child born in the usual way."

The liberal (or libertarian or liberationist) perspective sets cloning in the context of rights, freedoms and personal empowerment. Cloning is just a new option for exercising an individual's right to reproduce or to have the kind of child that he or she wants. Alternatively, cloning enhances our liberation (especially women's liberation) from the confines of nature, the vagaries of chance, or the necessity for sexual mating. Indeed, it liberates women from the need for men altogether, for the process requires only eggs, nuclei and (for the time being) uteri—plus, of course, a healthy dose of our (allegedly "masculine") manipulative science that likes to do all these things to mother nature and nature's mothers. For those who hold this outlook, the only moral restraints on cloning are adequately informed consent and the avoidance of bodily harm. If no one is cloned without her consent, and if the clonant is not physically damaged, then the liberal conditions for licit, hence moral, conduct are met. Worries that go beyond violating the will or maiming the body are dismissed as "symbolic"—which is to say, unreal.

The meliorist perspective embraces valetudinarians and also eugenicists. The latter were formerly more vocal in these discussions, but they are now generally happy to see their goals advanced under the less threatening banners of freedom and technological growth. These people see in cloning a new prospect for improving human beings—minimally, by ensuring the perpetuation of healthy individuals by avoiding the risks of genetic disease inherent in the lottery of sex, and maximally, by producing "optimum babies," preserving outstanding genetic material, and (with the help of soon-to-come techniques for precise genetic engineering) enhancing inborn human capacities on many fronts. Here the morality of cloning as a means is justified solely by the excellence of the end, that is, by the outstanding traits or individuals cloned—beauty, or brawn, or brains.

These three approaches, all quintessentially American and all perfectly fine in their places, are sorely wanting as approaches to human procreation. It is, to say the least, grossly distorting to view the wondrous mysteries of birth, renewal and individuality, and the deep meaning of parent—child relations, largely through the lens of our reductive science and its potent technologies. Similarly, considering reproduction (and the intimate relations of family life!) primarily under the political-legal, adversarial, and individualistic notion of rights can only undermine the private yet fundamentally social, cooperative and duty-laden character of childbearing, child rearing and their bond to the covenant of marriage. Seeking to escape entirely from nature (in order to satisfy a natural desire or a natural right to reproduce!) is self-contradictory in theory and self-alienating in practice. For we are erotic beings only because we are embodied beings, and not merely intellects and wills unfortunately imprisoned in our bodies. And, though health and fitness are clearly great goods, there is something deeply disquieting in looking on our prospective children as artful products perfectible by genetic engineering, increasingly held to our willfully imposed designs, specifications and margins of tolerable error.

The technical, liberal and meliorist approaches all ignore the deeper anthropological, social and, indeed, ontological meanings of bringing forth new life. To this more fitting and profound point of view, cloning shows itself to be a major alteration, indeed, a major violation, of our given nature as embodied, gendered and engendering beings-and of the social relations built on this natural ground. Once this perspective is recognized, the ethical judgment on cloning can no longer be reduced to a matter of motives and intentions, rights and freedoms, benefits and harms, or even means and ends. It must be regarded primarily as a matter of meaning: Is cloning a fulfillment of human begetting and belonging? Or is cloning rather, as I contend, their pollution and perversion? To pollution and perversion, the fitting response can only be horror and revulsion; and conversely, generalized horror and revulsion are prima facie evidence of foulness and violation. The burden of moral argument must fall entirely on those who want to declare the widespread repugnances of humankind to be mere timidity or superstition.

Yet repugnance need not stand naked before the bar of reason. The wisdom of our horror at human cloning can be partially articulated, even if this is finally one of those instances about which the heart has its reasons that reason cannot entirely know.

The Profundity of Sex

To see cloning in its proper context, we must begin not, as I did before, with laboratory technique, but with the anthropology—natural and social—of sexual reproduction.

Sexual reproduction—by which I mean the generation of new life from (exactly) two complementary elements, one female, one male, (usually) through coitus—is established (if that is the right term) not by human decision, culture or tradition, but by nature; it is the natural way of all mammalian reproduction. By nature, each child has two complementary biological progenitors. Each child thus stems from and unites exactly two lineages. In natural generation, moreover, the precise genetic constitution of the resulting offspring is determined by a combination of nature and chance, not by human design: each human child shares the common natural human species genotype, each child is genetically (equally) kin to each (both) parent(s), yet each child is also genetically unique.

These biological truths about our origins foretell deep truths about our identity and about our human condition altogether. Every one of us is at once equally human, equally enmeshed in a particular familial nexus of origin, and equally individuated in our trajectory from birth to death—and, if all goes well, equally capable (despite our mortality) of participating, with a complementary other, in the very same renewal of such human possibility through procreation. Though less momentous than our common humanity, our genetic individuality is not humanly trivial. It shows itself forth in our distinctive appearance through which we are everywhere recognized; it is revealed in our "signature" marks of fingerprints and our self-recognizing immune system; it symbolizes and foreshadows exactly the unique, never-to-be-repeated character of each human life.

Human societies virtually everywhere have structured child-rearing responsibilities and systems of identity and relationship on the bases of these deep natural facts of begetting. The mysterious yet ubiquitous "love of one's own" is everywhere culturally exploited, to make sure that children are not just produced but well cared for and to create for everyone clear ties of meaning, belonging and obligation. But it is wrong to treat such naturally rooted social practices as mere cultural constructs (like left- or right-driving, or like burying or cremating the dead) that we can alter with little human cost. What would kinship be without its clear natural grounding? And what would identity be without kinship? We must resist those who have begun to refer to sexual reproduction as the "traditional method of reproduction," who would have us regard as merely traditional, and by implication arbitrary, what is in truth not only natural but most certainly profound.

Asexual reproduction, which produces "single-parent" offspring, is a radical departure from the natural human way, confounding all normal understandings of father, mother, sibling, grandparent, etc., and all moral relations tied thereto. It becomes even more of a radical departure when the resulting offspring is a clone derived not from an embryo, but from a mature adult to whom the clone would be an identical twin; and when the process occurs not by natural accident (as in natural twinning), but by deliberate human design and manipulation; and when the child's (or children's) genetic constitution is pre-selected by the parent(s) (or scientists). Accordingly, as we will see, cloning is vulnerable to three kinds of concerns and objections, related to these three points: cloning threatens confusion of identity and individuality, even in small-scale cloning; cloning represents a giant step (though not the first one) toward transforming procreation into manufacture, that is, toward the increasing depersonalization of the process of generation and, increasingly, toward the "production" of human children as artifacts, products of human will and design (what others have called the problem of "commodification" of new life); and cloning—like other forms of eugenic engineering of the next generation—represents a form of despotism of the cloners over the cloned, and thus (even in benevolent cases) represents a blatant violation of the inner meaning of parent–child relations, of what it means to have a child, of what it means to say "yes" to our own demise and "replacement." Before turning to these specific ethical objections, let me test my claim of the profundity of the natural way by taking up a challenge recently posed by a friend. What if the given natural human way of reproduction were asexual, and we now had to deal with a new technological innovation—artificially induced sexual dimorphism and the fusing of complementary gametes—whose inventors argued that sexual reproduction promised all sorts of advantages, including hybrid vigor and the creation of greatly increased individuality? Would one then be forced to defend natural asexuality because it was natural? Could one claim that it carried deep human meaning?

The response to this challenge broaches the ontological meaning of sexual reproduction. For it is impossible, I submit, for there to have been human life—or even higher forms of animal life—in the absence of sexuality and sexual reproduction. We find asexual reproduction only in the lowest forms of life: bacteria, algae, fungi, some lower invertebrates. Sexuality brings with it a new and enriched relationship to the world. Only sexual animals can seek and find complementary others with whom to pursue a goal that transcends their own existence. For a sexual being, the world is no longer an indifferent and largely homogeneous otherness, in part edible, in part dangerous. It also contains some very special and related and complementary beings, of the same kind but of opposite sex, toward whom one reaches out with special interest and intensity. In higher birds and mammals, the outward gaze keeps a lookout not only for food and predators, but also for prospective mates; the beholding of the many-splendored world is suffused with desire for union, the animal antecedent of human eros and the germ of sociality. Not by accident is the human animal both the sexiest animal—whose females do not go into heat but are receptive throughout the estrous cycle and whose males must therefore have greater sexual appetite and energy in order to reproduce successfully—and also the most aspiring, the most social, the most open and the most intelligent animal.

The soul-elevating power of sexuality is, at bottom, rooted in its strange connection to mortality, which it simultaneously accepts and tries to overcome. Asexual reproduction may be seen as a continuation of the activity of self-preservation. When one organism buds or divides to become two, the original being is (doubly) preserved, and nothing dies. Sexuality, by contrast, means perishability and serves replacement; of the two that come together to generate one soon will die. Sexual desire, in human beings as in animals, thus serves an end that is partly hidden from, and finally at odds with, the self-serving individual. Whether we know it or not, when we are sexually active we are voting with our genitalia for our own demise. The salmon swimming upstream to spawn and die tell the universal story: sex is bound up with death, to which it holds a partial answer in procreation.

The salmon and the other animals evince this truth blindly. Only the human being can understand what it means. As we learn so powerfully from the story of the Garden of Eden, our humanization is coincident with sexual self-consciousness, with the recognition of our sexual nakedness and all that it implies: shame at our needy incompleteness, unruly self-division and finitude; awe before the eternal; hope in the self-transcending possibilities of children and a relationship to the divine. In the sexually self-conscious animal, sexual desire can become eros, lust can become love. Sexual desire humanly regarded is thus sublimated into erotic longing for wholeness, completion and immortality, which drives us knowingly into the embrace and its generative fruit—as well as into all the higher human possibilities of deed, speech and song.

Through children, a good common to both husband and wife, male and female achieve some genuine unification (beyond the mere sexual "union," which fails to do so). The two become one through sharing generous (not needy) love for this third being as good. Flesh of their flesh, the child is the parents' own commingled being externalized, and given a separate and persisting existence. Unification is enhanced also by their commingled work of rearing. Providing an opening to the future beyond the grave, carrying not only our seed but also our names, our ways and our hopes that they will surpass us in goodness and happiness, children are a testament to the possibility of transcendence. Gender duality and sexual desire, which first draws our love upward and outside of ourselves, finally provide for the partial overcoming of the confinement and limitation of perishable embodiment altogether.

Human procreation, in sum, is not simply an activity of our rational wills. It is a more complete activity precisely because it engages us bodily, erotically and spiritually, as well as rationally. There is wisdom in the mystery of nature that has joined the pleasure of sex, the inarticulate longing for union, the communication of the loving embrace and the deep-seated and only partly articulate desire for children in the very activity by which we continue the chain of human existence and participate in the renewal of human possibility. Whether or not we know it, the severing of procreation from sex, love and intimacy is inherently dehumanizing, no matter how good the product.

We are now ready for the more specific objections to cloning.

The Perversities of Cloning

First, an important if formal objection: any attempt to clone a human being would constitute an unethical experiment upon the resulting child-to-be. As the animal experiments (frog and sheep) indicate, there are grave risks of mishaps and deformities. Moreover, because of what cloning means, one cannot presume a future cloned child's consent to be a clone, even a healthy one. Thus, ethically speaking, we cannot even get to know whether or not human cloning is feasible.

I understand, of course, the philosophical difficulty of trying to compare a life with defects against nonexistence. Several bioethicists, proud of their philosophical cleverness, use this conundrum to embarrass claims that one can injure a child in its conception, precisely because it is only thanks to that complained-of conception that the child is alive to complain. But common sense tells us that we have no reason to fear such philosophisms. For we surely know that people can harm and even maim children in the very act of conceiving them, say, by paternal transmission of the aids virus, maternal transmission of heroin dependence or, arguably, even by bringing them into being as bastards or with no capacity or willingness to look after them properly. And we believe that to do this intentionally, or even negligently, is inexcusable and clearly unethical.

The objection about the impossibility of presuming consent may even go beyond the obvious and sufficient point that a clonant, were he subsequently to be asked, could rightly resent having been made a clone. At issue are not just benefits and harms, but doubts about the very independence needed to give proper (even retroactive) consent, that is, not just the capacity to choose but the disposition and ability to choose freely and well. It is not at all clear to what extent a clone will truly be a moral agent. For, as we shall see, in the very fact of cloning, and of rearing him as a clone, his makers subvert the cloned child's independence, beginning with that aspect that comes from knowing that one was an unbidden surprise, a gift, to the world, rather than the designed result of someone's artful project.

Cloning creates serious issues of identity and individuality. The cloned person may experience concerns about his distinctive identity not only because he will be in genotype and appearance identical to another human being, but, in this case, because he may also be twin to the person who is his "father" or "mother"—if one can still call them that. What would be the psychic burdens of being the "child" or "parent" of your twin? The cloned individual, moreover, will be saddled with a genotype that has already lived. He will not be fully a surprise to the world.

People are likely always to compare his performances in life with that of his alter ego. True, his nurture and his circumstance in life will be different; genotype is not exactly destiny. Still, one must also expect parental and other efforts to shape this new life after the original—or at least to view the child with the original version always firmly in mind. Why else did they clone from the star basketball player, mathematician and beauty queen—or even dear old dad—in the first place?

Since the birth of Dolly, there has been a fair amount of doublespeak on this matter of genetic identity. Experts have rushed in to reassure the public that the clone would in no way be the same person, or have any confusions about his or her identity: as previously noted, they are pleased to point out that the clone of Mel Gibson would not be Mel Gibson. Fair enough. But one is shortchanging the truth by emphasizing the additional importance of the intrauterine environment, rearing and social setting: genotype obviously matters plenty. That, after all, is the only reason to clone, whether human beings or sheep. The odds that clones of Wilt Chamberlain will play in the NBA are, I submit, infinitely greater than they are for clones of Robert Reich.

Curiously, this conclusion is supported, inadvertently, by the one ethical sticking point insisted on by friends of cloning: no cloning without the donor's consent. Though an orthodox liberal objection, it is in fact quite puzzling when it comes from people (such as Ruth Macklin) who also insist that genotype is not identity or individuality, and who deny that a child could reasonably complain about being made a genetic copy. If the clone of Mel Gibson would not be Mel Gibson, why should Mel Gibson have grounds to object that someone had been made his clone? We already allow researchers to use blood and tissue samples for research purposes of no benefit to their sources: my falling hair, my expectorations, my urine and even my biopsied tissues are "not me" and not mine. Courts have held that the profit gained from uses to which scientists put my discarded tissues do not legally belong to me. Why, then, no cloning without consent—including, I assume, no cloning from the body of someone who just died? What harm is done the donor, if genotype is "not me"? Truth to tell, the only powerful justification for objecting is that genotype really does have something to do with identity, and everybody knows it. If not, on what basis could Michael Jordan object that someone cloned "him," say, from cells taken from a "lost" scraped-off piece of his skin? The insistence on donor consent unwittingly reveals the problem of identity in all cloning.

Genetic distinctiveness not only symbolizes the uniqueness of each human life and the independence of its parents that each human child rightfully attains. It can also be an important support for living a worthy and dignified life. Such arguments apply with great force to any large-scale replication of human individuals. But they are sufficient, in my view, to rebut even the first attempts to clone a human being. One must never forget that these are human beings upon whom our eugenic or merely playful fantasies are to be enacted.

Troubled psychic identity (distinctiveness), based on all-too-evident genetic identity (sameness), will be made much worse by the utter confusion of social identity and kinship ties. For, as already noted, cloning radically confounds lineage and social relations, for "offspring" as for "parents." As bioethicist James Nelson has pointed out, a female child cloned from her "mother" might develop a desire for a relationship to her "father," and might understandably seek out the father of her "mother," who is after all also her biological twin sister. Would "grandpa," who thought his paternal duties were concluded, be pleased to discover that the clonant looked to him for paternal attention and support?

Social identity and social ties of relationship and responsibility are widely connected to, and supported by, biological kinship. Social taboos on incest (and adultery) everywhere serve to keep clear who is related to whom (and especially which child belongs to which parents), as well as to avoid confounding the social identity of parent and child (or brother and sister) with the social identity of lovers, spouses and co-parents. True, social identity is altered by adoption (but as a matter of the best interest of already living children: we do not deliberately produce children for adoption). True, artificial insemination and in vitro fertilization with donor sperm, or whole embryo donation, are in some way forms of "prenatal adoption"—a not altogether unproblematic practice. Even here, though, there is in each case (as in all sexual reproduction) a known male source of sperm and a known single female source of egg—agenetic father and a genetic mother—should anyone care to know (as adopted children often do) who is genetically related to whom.

In the case of cloning, however, there is but one "parent." The usually sad situation of the "single-parent child" is here deliberately planned, and with a vengeance. In the case of self-cloning, the "offspring" is, in addition, one's twin; and so the dreaded result of incest—to be parent to one's sibling—is here brought about deliberately, albeit without any act of coitus. Moreover, all other relationships will be confounded. What will father, grandfather, aunt, cousin, sister mean? Who will bear what ties and what burdens? What sort of social identity will someone have with one whole side—"father's" or "mother's"—necessarily excluded? It is no answer to say that our society, with its high incidence of divorce, remarriage, adoption, extramarital childbearing and the rest, already confounds lineage and confuses kinship and responsibility for children (and everyone else), unless one also wants to argue that this is, for children, a preferable state of affairs.

Human cloning would also represent a giant step toward turning begetting into making, procreation into manufacture (literally, something "handmade"), a process already begun with in vitro fertilization and genetic testing of embryos. With cloning, not only is the process in hand, but the total genetic blueprint of the cloned individual is selected and determined by the human artisans. To be sure, subsequent development will take place according to natural processes; and the resulting children will still be recognizably human. But we here would be taking a major step into making man himself simply another one of the man-made things. Human nature becomes merely the last part of nature to succumb to the technological project, which turns all of nature into raw material at human disposal, to be homogenized by our rationalized technique according to the subjective prejudices of the day. How does begetting differ from making? In natural procreation, human beings come together, complementarily male and female, to give existence to another being who is formed, exactly as we were, by what we are: living, hence perishable, hence aspiringly erotic, human beings. In clonal reproduction, by contrast, and in the more advanced forms of manufacture to which it leads, we give existence to a being not by what we are but by what we intend and design. As with any product of our making, no matter how excellent, the artificer stands above it, not as an equal but as a superior, transcending it by his will and creative prowess. Scientists who clone animals make it perfectly clear that they are engaged in instrumental making; the animals are, from the start, designed as means to serve rational human purposes. In human cloning, scientists and prospective "parents" would be adopting the same technocratic mentality to human children: human children would be their artifacts.

Such an arrangement is profoundly dehumanizing, no matter how good the product. Mass-scale cloning of the same individual makes the point vividly; but the violation of human equality, freedom, and dignity are present even in a single planned clone. And procreation dehumanized into manufacture is further degraded by commodification, a virtually inescapable result of allowing babymaking to proceed under the banner of commerce. Genetic and reproductive biotechnology companies are already growth industries, but they will go into commercial orbit once the Human Genome Project nears completion. Supply will create enormous demand. Even before the capacity for human cloning arrives, established companies will have invested in the harvesting of eggs from ovaries obtained at autopsy or through ovarian surgery, practiced embryonic genetic alteration, and initiated the stockpiling of prospective donor tissues. Through the rental of surrogate-womb services, and through the buying and selling of tissues and embryos, priced according to the merit of the donor, the commodification of nascent human life will be unstoppable.

Finally, and perhaps most important, the practice of human cloning by nuclear transfer—like other anticipated forms of genetic engineering of the next generation—would enshrine and aggravate a profound and mischievous misunderstanding of the meaning of having children and of the parent—child relationship. When a couple now chooses to procreate, the partners are saying yes to the emergence of new life in its novelty, saying yes not only to having a child but also, tacitly, to having whatever child this child turns out to be. In accepting our finitude and opening ourselves to our replacement, we are tacitly confessing the limits of our control. In this ubiquitous way of nature, embracing the future by procreating means precisely that we are relinquishing our grip, in the very activity of taking up our own share in what we hope will be the immortality of human life and the human species. This means that our children are not our children: they are not our property, not our possessions. Neither are they supposed to live our lives for us, or anyone else's life but their own. To be sure, we seek to guide them on their way, imparting to them not just life but nurturing, love, and a way of life; to be sure, they bear our hopes that they will live fine and flourishing lives, enabling us in small measure to transcend our own limitations. Still, their genetic distinctiveness and independence are the natural foreshadowing of the deep truth that they have their own and never-before-enacted life to live. They are sprung from a past, but they take an uncharted course into the future.

Much harm is already done by parents who try to live vicariously through their children. Children are sometimes compelled to fulfill the broken dreams of unhappy parents; John Doe Jr. or the III is under the burden of having to live up to his forebear's name. Still, if most parents have hopes for their children, cloning parents will have expectations. In cloning, such overbearing parents take at the start a decisive step that contradicts the entire meaning of the open and forward-looking nature of parent—child relations. The child is given a genotype that has already lived, with full expectation that this blueprint of a past life ought to be controlling of the life that is to come. Cloning is inherently despotic, for it seeks to make one's children (or someone else's children) after one's own image (or an image of one's choosing) and their future according to one's will. In some cases, the despotism may be mild and benevolent. In other cases, it will be mischievous and downright tyrannical. But despotism—the control of another through one's will—it inevitably will be.

Meeting Some Objections

The defenders of cloning, of course, are not wittingly friends of despotism. Indeed, they regard themselves mainly as friends of freedom: the freedom of individuals to reproduce, the freedom of scientists and inventors to discover and devise and to foster "progress" in genetic knowledge and technique. They want large-scale cloning only for animals, but they wish to preserve cloning as a human option for exercising our "right to reproduce"—our right to have children, and children with "desirable genes." As law professor John Robertson points out, under our "right to reproduce" we already practice early forms of unnatural, artificial and extramarital reproduction, and we already practice early forms of eugenic choice. For this reason, he argues, cloning is no big deal.

We have here a perfect example of the logic of the slippery slope, and the slippery way in which it already works in this area. Only a few years ago, slippery slope arguments were used to oppose artificial insemination and in vitro fertilization using unrelated sperm donors. Principles used to justify these practices, it was said, will be used to justify more artificial and more eugenic practices, including cloning. Not so, the defenders retorted, since we can make the necessary distinctions. And now, without even a gesture at making the necessary distinctions, the continuity of practice is held by itself to be justificatory.

The principle of reproductive freedom as currently enunciated by the proponents of cloning logically embraces the ethical acceptability of sliding down the entire rest of the slope—to producing children ectogenetically from sperm to term (should it become feasible) and to producing children whose entire genetic makeup will be the product of parental eugenic planning and choice. If reproductive freedom means the right to have a child of one's own choosing, by whatever means, it knows and accepts no limits.

But, far from being legitimated by a "right to reproduce," the emergence of techniques of assisted reproduction and genetic engineering should compel us to reconsider the meaning and limits of such a putative right. In truth, a "right to reproduce" has always been a peculiar and problematic notion. Rights generally belong to individuals, but this is a right which (before cloning) no one can exercise alone. Does the right then inhere only in couples? Only in married couples? Is it a (woman's) right to carry or deliver or a right (of one or more parents) to nurture and rear? Is it a right to have your own biological child? Is it a right only to attempt reproduction, or a right also to succeed? Is it a right to acquire the baby of one's choice?

The assertion of a negative "right to reproduce" certainly makes sense when it claims protection against state interference with procreative liberty, say, through a program of compulsory sterilization. But surely it cannot be the basis of a tort claim against nature, to be made good by technology, should free efforts at natural procreation fail.

Some insist that the right to reproduce embraces also the right against state interference with the free use of all technological means to obtain a child. Yet such a position cannot be sustained: for reasons having to do with the means employed, any community may rightfully prohibit surrogate pregnancy, or polygamy, or the sale of babies to infertile couples, without violating anyone's basic human "right to reproduce." When the exercise of a previously innocuous freedom now involves or impinges on troublesome practices that the original freedom never was intended to reach, the general presumption of liberty needs to be reconsidered.

We do indeed already practice negative eugenic selection, through genetic screening and prenatal diagnosis. Yet our practices are governed by a norm of health. We seek to prevent the birth of children who suffer from known (serious) genetic diseases. When and if gene therapy becomes possible, such diseases could then be treated, in utero or even before implantation—I have no ethical objection in principle to such a practice (though I have some practical worries), precisely because it serves the medical goal of healing existing individuals. But therapy, to be therapy, implies not only an existing "patient." It also implies a norm of health. In this respect, even germline gene "therapy," though practiced not on a human being but on egg and sperm, is less radical than cloning, which is in no way therapeutic. But once one blurs the distinction between health promotion and genetic enhancement, between so-called negative and positive eugenics, one opens the door to all future eugenic designs. "To make sure that a child will be healthy and have good chances in life": this is Robertson's principle, and owing to its latter clause it is an utterly elastic principle, with no boundaries. Being over eight feet tall will likely produce some very good chances in life, and so will having the looks of Marilyn Monroe, and so will a genius-level intelligence.

Proponents want us to believe that there are legitimate uses of cloning that can be distinguished from illegitimate uses, but by their own principles no such limits can be found. (Nor could any such limits be enforced in practice.) Reproductive freedom, as they understand it, is governed solely by the subjective wishes of the parents-to-be (plus the avoidance of bodily harm to the child). The sentimentally appealing case of the childless married couple is, on these grounds, indistinguishable from the case of an individual (married or not) who would like to clone someone famous or talented, living or dead. Further, the principle here endorsed justifies not only cloning but, indeed, all future artificial attempts to create (manufacture) "perfect" babies.

A concrete example will show how, in practice no less than in principle, the so-called innocent case will merge with, or even turn into, the more troubling ones. In practice, the eager parents-to-be will necessarily be subject to the tyranny of expertise. Consider an infertile married couple, she lacking eggs or he lacking sperm, that wants a child of their (genetic) own, and propose to clone either husband or wife. The scientist-physician (who is also co-owner of the cloning company) points out the likely difficulties—a cloned child is not really their (genetic) child, but the child of only one of them; this imbalance may produce strains on the marriage; the child might suffer identity confusion; there is a risk of perpetuating the cause of sterility; and so on—and he also points out the advantages of choosing a donor nucleus. Far better than a child of their own would be a child of their own choosing. Touting his own expertise in selecting healthy and talented donors, the doctor presents the couple with his latest catalog containing the pictures, the health records and the accomplishments of his stable of cloning donors, samples of whose tissues are in his deep freeze. Why not, dearly beloved, a more perfect baby?

The "perfect baby," of course, is the project not of the infertility doctors, but of the eugenic scientists and their supporters. For them, the paramount right is not the so-called right to reproduce but what biologist Bentley Glass called, a quarter of a century ago, "the right of every child to be born with a sound physical and mental constitution, based on a sound genotype...the inalienable right to a sound heritage." But to secure this right, and to achieve the requisite quality control over new human life, human conception and gestation will need to be brought fully into the bright light of the laboratory, beneath which it can be fertilized, nourished, pruned, weeded, watched, inspected, prodded, pinched, cajoled, injected, tested, rated, graded, approved, stamped, wrapped, sealed and delivered. There is no other way to produce the perfect baby.

Yet we are urged by proponents of cloning to forget about the science fiction scenarios of laboratory manufacture and multiple-copied clones, and to focus only on the homely cases of infertile couples exercising their reproductive rights. But why, if the single cases are so innocent, should multiplying their performance be so off-putting? (Similarly, why do others object to people making money off this practice, if the practice itself is perfectly acceptable?) When we follow the sound ethical principle of universalizing our choice—"would it be right if everyone cloned a Wilt Chamberlain (with his consent, of course)? Would it be right if everyone decided to practice asexual reproduction?"—we discover what is wrong with these seemingly innocent cases. The so-called science fiction cases make vivid the meaning of what looks to us, mistakenly, to be benign.

Though I recognize certain continuities between cloning and, say, in vitro fertilization, I believe that cloning differs in essential and important ways. Yet those who disagree should be reminded that the "continuity" argument cuts both ways. Sometimes we establish bad precedents, and discover that they were bad only when we follow their inexorable logic to places we never meant to go. Can the defenders of cloning show us today how, on their principles, we will be able to see producing babies ("perfect babies") entirely in the laboratory or exercising full control over their genotypes (including so-called enhancement) as ethically different, in any essential way, from present forms of assisted reproduction? Or are they willing to admit, despite their attachment to the principle of continuity, that the complete obliteration of "mother" or "father," the complete depersonalization of procreation, the complete manufacture of human beings and the complete genetic control of one generation over the next would be ethically problematic and essentially different from current forms of assisted reproduction? If so, where and how will they draw the line, and why? I draw it at cloning, for all the reasons given.

Ban the Cloning of Humans

What, then, should we do? We should declare that human cloning is unethical in itself and dangerous in its likely consequences. In so doing, we shall have the backing of the overwhelming majority of our fellow Americans, and of the human race, and (I believe) of most practicing scientists. Next, we should do all that we can to prevent the cloning of human beings. We should do this by means of an international legal ban if possible, and by a unilateral national ban, at a minimum. Scientists may secretly undertake to violate such a law, but they will be deterred by not being able to stand up proudly to claim the credit for their technological bravado and success. Such a ban on clonal babymaking, moreover, will not harm the progress of basic genetic science and technology. On the contrary, it will reassure the public that scientists are happy to proceed without violating the deep ethical norms and intuitions of the human community.

This still leaves the vexed question about laboratory research using early embryonic human clones, specially created only for such research purposes, with no intention to implant them into a uterus. There is no question that such research holds great promise for gaining fundamental knowledge about normal (and abnormal) differentiation, and for developing tissue lines for transplantation that might be used, say, in treating leukemia or in repairing brain or spinal cord injuries-to mention just a few of the conceivable benefits. Still, unrestricted clonal embryo research will surely make the production of living human clones much more likely. Once the genies put the cloned embryos into the bottles, who can strictly control where they go (especially in the absence of legal prohibitions against implanting them to produce a child)?

I appreciate the potentially great gains in scientific knowledge and medical treatment available from embryo research, especially with cloned embryos. At the same time, I have serious reservations about creating human embryos for the sole purpose of experimentation. There is something deeply repugnant and fundamentally transgressive about such a utilitarian treatment of prospective human life. This total, shameless exploitation is worse, in my opinion, than the "mere" destruction of nascent life. But I see no added objections, as a matter of principle, to creating and using cloned early embryos for research purposes, beyond the objections that I might raise to doing so with embryos produced sexually.

And yet, as a matter of policy and prudence, any opponent of the manufacture of cloned humans must, I think, in the end oppose also the creating of cloned human embryos. Frozen embryonic clones (belonging to whom?) can be shuttled around without detection. Commercial ventures in human cloning will be developed without adequate oversight. In order to build a fence around the law, prudence dictates that one oppose—for this reason alone—all production of cloned human embryos, even for research purposes. We should allow all cloning research on animals to go forward, but the only safe trench that we can dig across the slippery slope, I suspect, is to insist on the inviolable distinction between animal and human cloning.

Some readers, and certainly most scientists, will not accept such prudent restraints, since they desire the benefits of research. They will prefer, even in fear and trembling, to allow human embryo cloning research to go forward.

Very well. Let us test them. If the scientists want to be taken seriously on ethical grounds, they must at the very least agree that embryonic research may proceed if and only if it is preceded by an absolute and effective ban on all attempts to implant into a uterus a cloned human embryo (cloned from an adult) to produce a living child. Absolutely no permission for the former without the latter.

The National Bioethics Advisory Commission's recommendations regarding this matter should be watched with the greatest care. Yielding to the wishes of the scientists, the commission will almost surely recommend that cloning human embryos for research be permitted. To allay public concern, it will likely also call for a temporary moratorium—not a legislative ban—on implanting cloned embryos to make a child, at least until such time as cloning techniques will have been perfected and rendered "safe" (precisely through the permitted research with cloned embryos). But the call for a moratorium rather than a legal ban would be a moral and a practical failure. Morally, this ethics commission would (at best) be waffling on the main ethical question, by refusing to declare the production of human clones unethical (or ethical). Practically, a moratorium on implantation cannot provide even the minimum protection needed to prevent the production of cloned humans.

Opponents of cloning need therefore to be vigilant. Indeed, no one should be willing even to consider a recommendation to allow the embryo research to proceed unless it is accompanied by a call for prohibiting implantation and until steps are taken to make such a prohibition effective.

Technically, the National Bioethics Advisory Commission can advise the president only on federal policy, especially federal funding policy. But given the seriousness of the matter at hand, and the grave public concern that goes beyond federal funding, the commission should take a broader view. (If it doesn't, Congress surely will.) Given that most assisted reproduction occurs in the private sector, it would be cowardly and insufficient for the commission to say, simply, "no federal funding" for such practices. It would be disingenuous to argue that we should allow federal funding so that we would then be able to regulate the practice; the private sector will not be bound by such regulations. Far better, for virtually everyone concerned, would be to distinguish between research on embryos and baby-making, and to call for a complete national and international ban (effected by legislation and treaty) of the latter, while allowing the former to proceed (at least in private laboratories).

The proposal for such a legislative ban is without American precedent, at least in technological matters, though the British and others have banned cloning of human beings, and we ourselves ban incest, polygamy and other forms of "reproductive freedom." Needless to say, working out the details of such a ban, especially a global one, would be tricky, what with the need to develop appropriate sanctions for violators. Perhaps such a ban will prove ineffective; perhaps it will eventually be shown to have been a mistake. But it would at least place the burden of practical proof where it belongs: on the proponents of this horror, requiring them to show very clearly what great social or medical good can be had only by the cloning of human beings.

We Americans have lived by, and prospered under, a rosy optimism about scientific and technological progress. The technological imperative—if it can be done, it must be done—has probably served us well, though we should admit that there is no accurate method for weighing benefits and harms. Even when, as in the cases of environmental pollution, urban decay or the lingering deaths that are the unintended by-products of medical success, we recognize the unwelcome outcomes of technological advance, we remain confident in our ability to fix all the "bad" consequences—usually by means of still newer and better technologies. How successful we can continue to be in such post hoc repairing is at least an open question. But there is very good reason for shifting the paradigm around, at least regarding those technological interventions into the human body and mind that will surely effect fundamental (and likely irreversible) changes in human nature, basic human relationships, and what it means to be a human being. Here we surely should not be willing to risk everything in the naive hope that, should things go wrong, we can later set them right.

The president's call for a moratorium on human cloning has given us an important opportunity. In a truly unprecedented way, we can strike a blow for the human control of the technological project, for wisdom, prudence and human dignity. The prospect of human cloning, so repulsive to contemplate, is the occasion for deciding whether we shall be slaves of unregulated progress, and ultimately its artifacts, or whether we shall remain free human beings who guide our technique toward the enhancement of human dignity. If we are to seize the occasion, we must, as the late Paul Ramsey wrote, raise the ethical questions with a serious and not a frivolous conscience. A man of frivolous conscience announces that there are ethical quandaries ahead that we must urgently consider before the future catches up with us. By this he often means that we need to devise a new ethics that will provide the rationalization for doing in the future what men are bound to do because of new actions and interventions science will have made possible. In contrast a man of serious conscience means to say in raising urgent ethical questions that there may be some things that men should never do. The good things that men do can be made complete only by the things they refuse to do.

Is Biomedical Research Too Dangerous to Pursue?

Arthur Caplan

A particular ethical stance, utilitarianism, has guided science policy in the United States for more than 50 years (see references 1 and 2 below). The moral argument that investing in science and technology extends life and improves the quality of life, despite exacting a toll in harms and risks, has long dominated debate in American science policy. This ethical framework is now under sustained attack from many quarters (reff. 2–10), not because of doubts about the benefits scientific knowledge can bring, but because of the belief that important values will of necessity be compromised if biomedicine continues in its current direction (3–10). Some making this argument have the ears of those at the highest levels of government (7–9).

During the past year, a rash of writings has appeared that reject the view that benefit alone should determine whether to proceed with the biotechnology revolution. As the bioethicist Daniel Callahan argues, those concerned about where biotechnology may take us do not and should not accept the framing of the debate in utilitarian talk of "progress," "cures," and "a better life" (3). Worries about the loss of our humanity, the value of life, and the meaning of human experience are at the core of their fears.

Perhaps the most outspoken and influential critic of the utilitarian justification for research is the Chair of President Bush's Council on Bioethics, Leon R. Kass. Kass has long been concerned about the ways in which biotechnology undermines or shifts our understanding of the nature of family, marriage, sexual relations, aging, and parenting (11). Recent developments in cloning, stem cells, genetic testing, pharmacology, anti-aging research, and the neurosciences have only intensified his concerns.

Appeals to a utilitarian ethic—the benefits to be gained from biomedical research—have been the overarching ethical ethos grounding U.S. science policy. But, the suggestion that science policy in the United States has not recognized limits to utilitarianism is a straw man. Patient and subject rights, respect for privacy, and the need to have research reviewed by appropriate third parties are examples of such limits. Current critics want to go far beyond limits, in some cases, to outright bans or prohibitions. The case they offer is far from persuasive.

Their moral worries fall into three areas (7–10). First, biomedical research cannot continue on its present course without significantly altering human nature. Second, if, in the name of more cures, longer life, and improved quality of life, we continue on our present biomedical research course, we will commodify and objectify human life (7, 10). Third, too much biomedical tinkering will produce a loss of authenticity and meaning in human experience (4, 6). We may someday feel better because of powerful drugs or immersion in a world of computer-generated stimuli, but we will be far less healthy because our sense of well-being will have become programmed, artificial, and inauthentic.

Must progress in biomedicine distort who we are? Has a case been made to slow or stop the biomedical research enterprise? I believe not.

Biomedical knowledge may lead us to tinker with the genes, neurons, or physical bodies that we understand today as essential to defining human nature. But human nature itself has changed drastically in response to technology. Even the most basic ideas about who we are—how we see the world; how we walk, run, and move; whom we interact with, befriend, and love; what thrills us and threatens us—are all the result of complex interactions between the world, technology, and our bodies (12). Nor is there any reason to glorify a particular phase in the evolution of human nature and declare it sacrosanct. Human nature is not static; it lacks any recognized "essence" and has elements that have proven maladaptive in the past.

Similarly, although we may imperil the value of humanity by seeing ourselves as portable sources of marketable organs or by using technology in order to ensure optimal reproductive success, choosing such objectified or commodified visions is not an inevitable result of biomedical progress. Social and political choices, not scientific advances, will determine how our dignity and autonomy are to be squared with the prospect of birthing from artificial wombs or enhancing our mental capacities through genetic or neuronal engineering.

Self-esteem need not be a victim of progress. Those born as a result of the application of forceps, neonatal intensive care units, in vitro fertilization, or preimplantation genetic diagnosis do not appear to suffer from undue angst about having been artificially "manufactured." Nor should they. There is not anything obviously more dignified about being "made" in the back seat of a car than there is having been conceived from a pipette and a sperm donor.

Finally, the concern that advances in biotechnology will come at a terrible price—the loss of authentic happiness, the loss of what makes life meaningful—struggle, suffering, frailty, finitude, and death (4, 6, 7, 9, 10) do not seem to square with what we have already experienced in the wake of biomedical progress. Do those who use glasses. insulin injections. wheelchairs, inhalers, oxygen tanks, hearing aids, or prosthetic limbs feel inauthentic or overcome by a loss of meaning in their lives? If I use a calculator, a computer. or the Internet to solve a problem, do I feel that I have been cheated out of a more authentic experience enjoyed by my grandparents, who used pen and paper calculation, visited a library, or mastered the multiplication table? There is little evidence for the dour view that we can only be happy when we have earned our happiness.

When the stakes are enormous—continued premature death, disability, chronic suffering—then much more is required of those who would challenge the wisdom of the aggressive pursuit of biomedical knowledge that is the only hope of solving these terrible problems.

References and Notes

1. B. L. R. Smith, *American Science Policy Since World WarII* (Brookings Institution, Washington, DC, 1990).
2. D. S. Greenberg, *Science. Money. and Politics* (Univ. of Chicago Press, Chicago 2001).
3. D. Callahan, *What Price Better Health?* (Univ. of California Press, Berkeley, CA 2003).
4. C. Elliott, *Better Than Well* (Norton, New York 2003)
5. S. S. Krimsky, *Science in the Private Interest* (Rowman & Littlefield, Lanham, MD 2003).
6. B. McKibben, *Enough: Staying Human in an Engineered Age* (Times Books, New York 2003).
7. L. Kass (President's Council on Bioethics), *Beyond Therapy: Biotechnology and the Pursuit of Happiness* (HarperCollins, New York 2003).
8. A. Wolfson, *New Atlantis* 2003 (no. 2), 55 (2003).
9. W. Kristol and E. Cohen, eds., *The Future Is Now: America Confronts the New Genetics* (Rowman & Littlefield, London 2002).
10. F. Fukuyama, *Our Posthuman Nature* (Picador, New York 2002).
11. R. Kass, *Public Interest* 26, 18 (1972).
12. E. Tanner, *Our Own Devices* (Knopf, New York 2003).

Cloning as a Reproductive Right

John Robertson

A proper assessment of human cloning requires that it be viewed in light of how it might be actually used once it is shown to be safe and effective. The most likely uses would involve extending current reproductive and genetic selection technologies. Several plausible uses can be articulated, quite different from the horrific scenarios currently imagined. The question becomes: Do these uses fall within mainstream understandings of why procreative freedom warrants special respect as one of our fundamental liberties? Investigation of this question will set the stage for examining what public policy toward human cloning ought to be.

The Demand for Human Cloning

Legitimate, family-centered uses of cloning are likely precisely because cloning is above all a commitment to have and rear a child. It will involve obtaining eggs, acquiring the DNA to be cloned, transfer of that DNA to a denucleated egg, placement of the activated embryo in a uterus, gestation, and the nurturing and rearing that the birth of any child requires. In addition, it will require a psychological commitment and ability to deal with the novelty of raising a child whose genome has been chosen, and who may be the later-born identical twin of another person, living or dead.

The most bizarre or horrific scenarios of cloning conveniently overlook the basic reality that human cloning requires a gestating uterus and a commitment to rear. The gestating mother is eliminated through the idea of total laboratory gestation as imagined in Huxley's *Brave New World*, or through high-tech surgery as in the movie *Multiplicity*. In others scenarios, it is thought that an evildoer can hire several women to gestate copies, with little thought given to how they would be reared or molded to be like the clone source. Nearly all of them overlook the impact of environmental influences on the cloned child, and the duties and burdens that rearing any child requires. They also overlook the extent to which the cloned child is not the property or slave of the initiator, but a person in her own right with all the rights and duties of other persons.

Because cloning is first and foremost the commitment to have and rear a child, it is most likely to appeal to those who wish to have a family but cannot easily do so by ordinary coital means. In some cases, they would turn to cloning because of the advantages that it offers over other assisted reproductive techniques. Or they would choose cloning because they have a need to exercise genetic choice over offspring, as in the desire for a healthy child or for a child to serve as a tissue donor. Given the desire to have healthy children, it is unlikely that couples will be interested in cloning unless they have good reason for thinking that the procedure is safe and effective, and that only healthy children will be born.

The question for moral, legal, and policy analysis is to assess the needs that such uses serve—both those expected to be typical and those that seem more bizarre—and their importance relative to other reproductive and genetic selection endeavors. Can cloning be used responsibly to help a couple achieve legitimate reproductive or family formation goals? If so, are these uses properly characterized as falling within the procreative liberty of individuals, and thus not subject to state restriction without proof of compelling harm?

To assess these questions, we must first investigate the meaning of procreative liberty, and then ask whether uses of cloning to enhance fertility, substitute for a gamete or embryo donor, produce organs or tissue for transplant, or pick a particular genome fall within common understandings of that liberty.

Human Cloning and Procreative Liberty

Procreative liberty is the freedom to decide whether or not to have offspring. It is a deeply accepted moral value, and pervades many of our social practices.[1] Its importance stems from the impact which having or not having offspring has in our lives. This is evident in the case of a choice to avoid reproduction. Because reproduction imposes enormous physical burdens for the woman, as well as social, psychological, and emotional burdens on both men and women, it is widely thought that people should not have to bear those burdens unless they voluntarily choose to.

But the desire to reproduce is also important. It connects people with nature and the next generation, gives them a sense of immortality, and enables them to rear and parent children. Depriving a person of the ability or opportunity to reproduce is a major burden and also should not occur without their consent.

Reproductive freedom—the freedom to decide whether or not to have offspring—is generally thought to be an important instance of personal liberty. Indeed, given its great impact on a person, it is considered a fundamental personal liberty. In recent years the emergence of assisted reproduction, noncoital means of conception, and prebirth genetic selection has also raised controversies about the limits of procreative freedom. The question of whether cloning is part of procreative liberty is a serious one only if noncoital, assisted reproduction and genetic selection are themselves part of that liberty. A strong argument exists that the moral right to reproduce does include the right to use noncoital or assisted means of reproduction. Infertile couples have the same interests in reproducing as coitally fertile couples, and the same abilities to rear children. That they are coitally infertile should no more bar them from reproducing with technical assistance than visual blindness should bar a person from reading with Braille or the aid of a reader. It thus follows that married couples (and arguably single persons as well) have a moral right to use noncoital assisted reproductive techniques, such as in vitro fertilization (IVF) and artificial insemination with spouse or partner's sperm, to beget biologically related offspring for rearing. It should also follow—though this is more controversial—that the infertile couple would have the right to use gamete donors, gestational surrogates, and even embryo donors if necessary. Although third-party collaborative reproduction does not replicate exactly the genes, gestation, and rearing unity that ordinarily arises in coital reproduction, they come very close and should be treated accordingly. Each of them, with varying degrees of closeness, enables the couple to have or rear children biologically related to at least one of them.

Some right to engage in genetic selection would also seem to follow from the right to decide whether or not to procreate.[2] People make decisions to reproduce or not because of the package of experiences that they think reproduction or its absence would bring. In many cases, they would not reproduce if it would lead to a packet of experiences X, but they would if it would produce packet Y. Since the makeup of the packet will determine whether or not they reproduce, a right to make reproductive decisions based on that packet should follow. Some right to choose characteristics, either by negative exclusion or positive selection, should follow as well, for the decision to reproduce may often depend upon whether the child will have the characteristics of concern.

If most current forms of assisted reproduction and genetic selection fall within prevailing notions of procreative freedom, then a strong argument exists that some forms of cloning are aspects of procreative liberty as well. For cloning shares many features with assisted reproduction and genetic selection, though there are also important differences. For example, the most likely uses of cloning would enable a married couple, usually infertile, to have healthy, biologically related children for rearing.[3] Or it would enable them to obtain a source of tissue for transplant to enable an existing child to live.

Cloning, however, is also different in important respects. Unlike the various forms of assisted reproduction, cloning is concerned not merely with producing a child, but rather with the genes that the resulting child will have. Many prebirth genetic selection techniques are now in wide use, but they operate negatively by excluding undesirable genetic characteristics, rather than positively, as cloning does. Moreover, none of them are able to select the entire nuclear genome of a child as cloning does.[4]

DNA Sources and Procreative Liberty

To assess whether cloning is protected as part of a married couple's procreative liberty, we must examine the several situations in which they might use nuclear transfer cloning to form a family. This will require addressing both the reasons or motivations driving a couple to clone, and the source of the DNA that they select for replication. It argues that cloning embryos, children, third parties, self, mate, or parents is an activity so similar to coital and noncoital forms of reproduction and family formation that they should be treated equivalently.

a. Cloning a couple's embryos. Cloning embryos, either by embryo splitting or nuclear transfer, would appear to be closely connected to procreative liberty. It is intended to enable a couple to rear a child biologically related to each, either by increasing the number of embryos available for transfer or by reducing the need to go through later IVF cycles.[5] Its reproductive status is clear whether the motivation for transfer of cloned embryos to the wife's uterus is simply to have another child or to replace a child who has died. In either case, transfer leads to the birth of a child from the couple's egg and sperm that they will rear.

Eventually, couples might seek to clone embryos, not to produce a child for rearing, but to produce an embryo from which tissue stem cells can be obtained for an existing child. In that case, cloning will not lead to the birth of another child. However, it involves use of their reproductive capacity. It may also enable existing children to live. It too should be found to be within the procreative or family autonomy of the couple.

b. Cloning one's children. The use of DNA from existing children to produce another child would also seem to fall within a couple's procreative liberty. This action is directly procreative because it leads to birth of a child who is formed from the egg and sperm of each spouse, even though it occurs asexually with the DNA of an existing child and not from a new union of egg and sperm.[6] Although it is novel to create a twin after one has already been born, it is still reproductive for the couple. The distinctly reproductive nature of their action is reinforced by the fact that they will gestate and rear the child that they clone.

The idea of cloning any existing child is plausibly foreseeable in several circumstances. One is where the parents want another child, and are so delighted with the existing one, that they simply want to create a twin of her, rather than take a chance on the genetic lottery.[7] A second is where an existing child might need an organ or tissue transplant. A third scenario would be to clone an existing child who has died, so that it might continue to live in another form with the parents.

The parental motivations in these cases are similar to parental motivations in coital reproduction. No one condemn parents who reproduce because they wanted a child as lovely as the first, they thought that a new child might be a tissue source for an existing child, or because they wanted another child after an earlier one had died. Given that the new child is cloned from the DNA of one of their own children, cloning one's own embryos or children to achieve those goals should also be regarded as an exercise of procreative liberty that deserves the special respect usually accorded to procreative choice.

c. Cloning third parties. A couple that seeks to use the DNA of a third party should also be viewed as forming a family in a way similar to family formation through coital conception. The DNA of another might be sought in lieu of gamete or embryo donation, though it could be chosen because of the source's characteristics or special meaning for the couple. The idea of "procreating" with the DNA of another raises several questions about the meaning and scope of procreative liberty that requires a more extended analysis. I begin with the initiating couple's rights or interests, and then ask whether the DNA source or her parents also have procreative interests and rights at stake.

(i) The initiating couple's cloning rights. We now ask whether the rights of procreation and family formation of a couple seeking to clone another's DNA would extend to use of a third party's DNA to create a child. Whether rearing is also intended turns out to be a key distinction. Let us first con-

sider the situation where rearing is intended, and then the situation where no rearing is intended.

A strong case for a right to clone another person (assuming that that person and her parents have consented) is where a married couple seeks to clone in order to have a child to rear. Stronger still is the case in which the wife will gestate the resulting embryo and commits to rearing the child. Is this an exercise of their procreative liberty that deserves the special protection that procreative choice generally receives?

The most likely requests to clone a third party would arise from couples that are not reasonably able to reproduce in other ways. A common situation would be where the couple both lack viable gametes, though the wife can gestate, and thus are candidates for embryo donation. Rather than obtain an embryo generated by an unknown infertile couple, through cloning they could choose the genotype of the child they will carry and rear.

Whose DNA might they choose? The DNA could be obtained from a friend or family member, although parents pose a problem. It could come from a sperm, egg, or DNA bank that provides DNA for a fee. Perhaps famous people would be willing to part with their DNA, in the same way that a sperm bank from Nobel Prize winners was once created. Rather than choose themselves, the couple might delegate the task of choosing the genome to their doctor or to some other party.

A strong argument exists that a couple using the DNA of a third party in lieu of embryo donation is engaged in a legitimate exercise of procreative liberty. The argument rests on the view that embryo donation is a protected part of that liberty. That view rests in turn on the recognition of the right of infertile couples to use gamete donation to form a family. As we have seen, coital infertility alone does not deprive one of the right to reproduce. Infertile persons have the same interests in having and rearing offspring and are as well equipped to rear as fertile couples. If that is so, the state could not ban infertile couples from using noncoital techniques to have children without a showing of tangible or compelling harm.

The preceding discussion is premised on the initiating couple's willingness to raise the cloned child. If no rearing is intended, however, their claim to clone the DNA of another should not be recognized as part of their procreative liberty, whether or not they have other means to reproduce. In this scenario an initiator procures the DNA and denucleated eggs, has embryos created from the DNA, and then either sells or provides the embryos to others, or commissions surrogates to gestate the embryos. The resulting children are then reared by others.

The crucial difference here is absence of intention to rear. If one is not intending to rear, then one's claim to be exercising procreative choice is much less persuasive. One is not directly reproducing because one's genes are not involved—they are not even being replicated. Nor is one gestating or rearing. Indeed, such a practice has many of the characteristics that made human cloning initially appear to be such a frightening proposition. It seems to treat children like fungible commodities produced for profit without regard to their well-being. It should not be deemed part of the initiating couple's procreative liberty.

(ii) The clone source's right to be cloned. Mention of the procreative rights of the person providing the DNA to be cloned is also relevant. I will assume that the clone source has consented to use of their DNA. If so, does she independently have the right to have a later-born genetic twin, such that a ban on cloning would violate her procreative rights as well?

The fact that she will not be rearing the child is crucial. Her claim to be cloned then is simply a claim to have another person exist in the world with her DNA (and note that if anonymity holds, she may never learn that her clone was born, much less ever meet her). If so, the only interest at stake is her interest in possibly having her DNA replicated without her gestation, rearing, or even perhaps her knowledge that a twin has been born. It is difficult to argue that this is a strong procreative interest, if it is a procreative interest at all. Thus, unless she undertook to rear the resulting child, the clone source would not have a fundamental right to be cloned.

But this assumes that being cloned is not itself reproduction. One could argue that cloning is quintessentially reproductive for the clone source because her entire genome is replicated. In providing the DNA for another child, she will be continuing her DNA into another generation. Given that the goal of sexual and asexual reproduction is the same—the continuation of one's DNA—and that individuals who are cloned might find or view it as a way of maintaining continuity with nature, we could plausibly choose to consider it a form of reproduction.

But even if we view the clone source as fully and clearly reproducing, she still is not rearing. Her claim of a right to be cloned is still a claim of a right to reproduction *tout court*—the barest and least protected form of reproduction.[8] If there is no rearing or gestation her claim merely to have her genes replicated will not qualify for moral or legal rights protection. If the clone source has a right here to be cloned, it would have to be derivative of the initiators' right to select source DNA for the child whom they will rear.

d. Cloning oneself. Another likely cloning scenario will involve cloning oneself. A strong case can be made that the use of one's own DNA to have and rear a child should be protected by procreative liberty.

As we have just seen, the right to clone oneself is weakest if rearing is not intended. Even if we grant that self-cloning is in some sense truly reproductive, rather than merely replicative, it would still be reproduction *tout court*, the minimal and least protected form of reproduction. Thus cloning oneself with no rearing intended would have no independent claim to be an exercise of procreative liberty. If it is protected at all, it would be derivative of the couple who then gestate and rear the cloned child.

The claim of a right to clone oneself is different if one plans to gestate and rear the resulting child. The situation is best viewed as a joint endeavor of the couple. As in embryo donation, the couple would gestate and rear. In this case, however, there would be a genetic relation between one of the rearing partners and the child—the relation of later-born identical twin.

The idea of a right to parent one's own later-born sibling is also plausibly viewed as a variation on the right to use a gamete donor. If such a right exists, it plausibly follows that that they would have the right to choose the gametes or gamete source they wish to use. A right to use their own DNA to have a child whom they then rear should follow. Using their own DNA has distinct advantages over the gametes purchased from commercial sperm banks or paid egg donors. One is more clearly continuing her own genetic line, one knows the gene source, and one is not buying gametes. Some persons might plausibly insist that they would have a family only if they could clone one of themselves because they are leery of the gametes of anonymous strangers.[9]

One might also argue for a right to have and rear one's own identical twin—the right to clone oneself—as a direct exercise of the right to reproduce. If one is free to reproduce, then one should also have the right to be cloned, because the genetic replication involved in cloning is directly and quintessentially reproductive. Duplicating one's entire DNA by nuclear cell transfer enables a person to survive longer than if cloning did not occur. To use Richard Dawkins' evocative term, the selfish gene wants to survive as long as possible, and will settle for cloning if that will do the job.[10] If rearing is intended, a person's procreative liberty should include the right to clone oneself. Only tangible harm to the child or others would then justify restrictions on self-cloning.

Constituting Procreative Liberty

The analysis has produced plausible arguments for finding that cloning is directly involved with procreative liberty in situations where the couple initiating the cloning intends to rear the resulting child. This protected interest is perhaps clearest when they are splitting embryos or using DNA from their own embryos or children, but it also holds when one of the rearing partner's DNA is used. Using the DNA of another person is less directly reproductive, but still maintains a gestational connection between the cloned child and its rearing parents, as now occurs in embryo donation.

In considering the relation between cloning and procreative liberty, we see once again how blurred the meanings of reproduction, family, parenting, and children become as we move away from sexual reproduction involving a couple's egg and sperm. Blurred meanings, however, can be clarified. The test must be how closely the marginal or deviant case is connected with the core. On this test, several plausible cases of a couple seeking to clone the DNA of embryos, children, themselves, or consenting third parties, can be articulated. In all instances they will be seeking a child whom they will gestate and rear. We do no great violence to prevailing understandings of procreative choice when we recognize DNA cloning to produce children whom we will rear as a legitimate form of family or procreative choice. Unless all selection is to be removed from reproduction, their interest in selecting the genes of their children deserves the same protection accorded other reproductive choices.

Public Policy

Having discussed the scientific questions and social controversies surrounding cloning, and the likely demand for it once it is shown to be safe and effective, we are now in a position to discuss public policy for human cloning. In formulating policy, however, we must take account of the state of the cloning art. One set of policy options applies when human applications are still in the research and development or experimental stage. Another set exists when research shows that human cloning is safe and effective. The birth of a sheep clone after 277 tries at somatic cell nuclear transfer has shown that much more research is needed before somatic cell cloning by nuclear transfer will be routinely available in sheep and other species, much less in humans. But an important set of policy issues will arise if animal and laboratory research shows that cloning is safe and effective in humans. Should all cloning then be permitted? Should some types of cloning be prohibited? What regulations will minimize the harms that cloning could cause?

Based on the analysis in this article, a ban on all human cloning, including the family-centered uses described above, is overbroad. But we must also ask if some uses of cloning should be forbidden and whether some regulation of permitted uses is desirable once human cloning becomes medically safe and feasible.

No cloning without rearing. A ban on human cloning unless the parties requesting the cloning will also rear is a much better policy than a ban on all cloning. The requirement of having to rear the clone addresses the worse abuses of cloning. It prevents a person from creating clones to be used as subjects or workers without regard for their own interests. For example, situations like that in *Boys from Brazil* or *Brave New World* would be prohibited, because the initiator is not rearing. This rule will assure the child a two-parent rearing situation—a prime determinant of a child's welfare. Furthermore, the rule would not violate the initiator's procreative liberty because merely producing children for others to rear is not an exercise of that liberty.

Ensuring that the initiating couple rears the child given the DNA of another prevents some risks to the child, but still leaves open the threats to individuality, autonomy, and kinship that many persons think that cloning presents. I have argued that parents who intend to have and rear a healthy child might not be as prey to those concerns as feared, yet some cloning situations, because of the novelty of choosing a genome, might still produce social or psychological problems.

Those risks should be addressed in terms of the situations most likely to generate them, and the regulations, short of prohibition, that might minimize their occurrence. It hardly follows that all cloning should be banned, because some undesirable cloning situations might occur. Like other slippery slope arguments, there is no showing that the bad uses are so likely to occur, or that if they did, their bad effects would so clearly outweigh the good, that one is justified in suffering the loss of the good to prevent the bad.

Notes

[1] The moral and legal arguments for procreative liberty are presented in John A. Robertson, *Children of Choice: Freedom and the New Reproductive Technologies* (Princeton 1994), pp. 22–42.

[2] For elaboration of this argument, see John A. Robertson, "Genetic Selection of Offspring Characteristics," 76 *B.U.L. Rev.* 421, 424–432 (1996).

[3] The article emphasizes the rights of married couples because they will be perceived as having a stronger claim to have children than unmarried persons. If their rights to clone are recognized, then the claims of unmarried persons to clone might follow.

[4] Since only nuclear DNA is transferred in cloning, DNA contained in the egg's cytoplasm in the form of mitochondria is not cloned or replicated (it is in the case of cloning by embryo splitting). The resulting child is thus not a true clone, for its mitochondrial DNA will have come from the egg source, who

will not usually also be providing the nucleus for transfer. Mitochondrial DNA is only a small portion of total DNA, perhaps 5%. However, malfunctions in it can still cause serious disease. See Douglas C. Wallace, "Mitochondrial DNA in Aging and Disease," 277 *Scientific American* 40 (August 1997).

[5] It might also be done to provide an embryo or child from whom tissue or organs for transplant for an existing child.

[6] Again, it might be used to create an embryo or child from whom tissue or organs for transplant into an existing child might be obtained.

[7] I am grateful to my colleague Charles Silver for this suggestion. However, other colleagues with children inform me that they would not clone an existing child, because they would want to see how the next child would differ.

[8] Reproduction *tout court* (without more) refers to genetic transmission without any rearing rights or duties in the resulting child, and in some cases, not even knowledge that a child has been born. The courts have not yet determined whether engaging in or avoiding reproduction *tout court* deserves the same protection that more robust or involved forms of reproduction have. For further discussion, see *Children of Choice*, pp. 108–109.

[9] Of course, this means that the other partner will have no DNA connection with the child she rears, unless she also provides the egg and mitochondria.

[10] In the long run, cloning might not be adaptive, because genetic diversity is needed. However, if the alternative is no genetic continuation at all—say because no reproduction occurs, or a gamete donor is chosen—then cloning increases the chance of long-term survival of the cloned DNA more than no cloning at all. Richard Dawkins would clone himself purely out of curiosity: "I find it a personally riveting thought that I could watch a small copy of myself nurtured through the early decades of the twenty-first century." (Peter Steinfels, "Beliefs," *New York Times*, July 12, 1997, p. A8).

From Regenerative Medicine to Human Design: What Are We Really Afraid Of?

Gregory Stock

Many of the public figures trying to shape policy in the life sciences these days—from Francis Fukuyama and Leon Kass to the perennial Jeremy Rifkin—are troubled by recent advances in biotechnology. They are not alone in their fears of the new possibilities that are emerging, their angst is shared even by some of the scientists at the vanguard of this research. As we push further into uncharted territory by deciphering and laying bare the workings of life, it is worth asking just what it is that so worries us.

The enormity of coming developments in molecular biology seems obvious, but their magnitude doesn't require that we respond with fear. We are hardly the first to appreciate the prospects for impending revolutionary developments in science. In 1780, Benjamin Franklin showed a very different attitude, when he wrote to the great English chemist Joseph Priestley, "The rapid progress true science makes occasions my regretting sometimes that I was born so soon. It is impossible to imagine the heights to which may be carried, in a thousand years, the power of man over matter."

Franklin was less bothered by impending changes from the next millennium of scientific discovery than by not being around to witness these amazing possibilities. Today, at the comparatively short remove of 225 years from that letter, the thousand-year span of his forecast seems conservative. But if Franklin could come back and see the extraordinary technologies that have arisen since his death, I suspect it would please him no end.

When we look at the possibilities embodied in the Human Genome Project, which are emblematic of those in proteomics, systems biology, and molecular biology in general, we see that this research is poised to carry humanity to destinations of new imagination.

The possibilities a mere century hence are so mind boggling that many scientists today who don't even believe in God are resorting to religious metaphor to try to communicate what is happening. Three years ago, the announcement of the rough sequence of the human genome provoked widespread commentary about finding the "Holy Grail of Biology," reading the "Book of Life," and breaking the "Code of Codes." And this unalloyed enthusiasm coupled with projections of rapid progress in biomedicine created soaring biotech valuations until investors began to realize how long and arduous the path might be from identifying gene targets to getting practical results and moving them into the clinic. The world has not yet shifted beneath our feet. Cancer and heart disease are still the biggest killers. We still get old and suffer the same diseases of aging. The promises of new drugs, gene therapies, and tissue engineering remain unfulfilled.

The excitement about the human genome was reminiscent of 1969, when Neil Armstrong walked on the moon, and we were about to race out towards the stars—a vision so perfectly embodied in Stanley Kubrick's classic film *2001: A Space Odyssey*. But that date has come and gone, and there is no HAL, and no space odyssey to our own moon, much less the moons of Jupiter. It is hardly surprising that people are wondering whether thirty-five years from now, our children will look back at this moment too and smile about the enthusiasm for regenerative medicine, life extension and designer babies. Will this all then seem like a crazy dream that brought us precious little?

I think not. Genetics and biology are at our core. And as we learn to adjust and modify these realms, we are learning to change ourselves. We have already used technology to transform the world around us. The canyons of glass, concrete, and stainless steel in any major city are not the stomping ground of our Pleistocene ancestors. And now our technology is becoming so potent and so precise that we are turning it back on our own selves. Before we're done, we will likely transform our own biology as much as we have already changed the world around us.

The sense that humanity is at the threshold of reworking its own biology is what troubles so many people. Clearly, medicine and healthcare will be transformed in the process. But these technologies will do much more than that. They will change the way we have children, alter how we manage our emotions, and even modify the human lifespan. Far sooner than people imagine, these core technologies will take us to the very question of what it means to be human.

Research in genomics, proteomics, genetic engineering, and regenerative medicine will be at the heart of these developments, and they will be subject not only to fickle public enthusiasm and angst, but to a tide of regulation, litigation, and political conflict.

The Big Picture

Two unprecedented revolutions are underway today. The

first is the silicon revolution: the telecommunications, computers, artificial intelligence, expert systems, and the related technology that is ever more shaping our lives. We have begun to breathe into inert sand—the silicon—at our feet a level of complexity rivaling life itself. And our world will never be the same.

The second revolution, a child of the first, is the one in molecular biology. As we plumb the workings of life and unravel our biology, we are seizing control of our evolutionary future. Our science has slammed evolution into "fast forward," and no one can say where the process will ultimately carry us. The only reason the world has not yet shifted beneath our feet is that we have barely begun. Even with all the world's seeming changes, today is but the calm before the storm.

The human genome project is a good example. We have a list of the human genes. We have libraries of their common variants. We have micro-array technology to screen for these variants at ever lower prices. And we are developing the bioinformatics capabilities to make sense of the Tsunami of such information about to sweep over us.

The impact of these developments will be enormous, because genes do matter. They are not our destiny, but they carry tremendous information about our predispositions and vulnerabilities in life, and about who we are. As a good rule of thumb—and this information comes from twin studies, not genomics—with adequate childhood environments such as those generally found in today's developed world, between a quarter and three quarters of the variation in most traits across the human population can be explained by genetics. For most of the traits we see as critical to our identities—our temperament, whether or not we are risk takers, aspects of our personality, our mental capabilities—our genes are a huge window into who we are. And now, we are drawing back the curtain.

Near-term Developments

The greatest immediate consequences of genomics in our lives will flow from the widespread, inexpensive, reliable personal genetic testing that will arrive over the next five to fifteen years. Such information may prove extremely challenging to us, both as individuals and as a society, but it hardly warrants the deep angst about biotechnology evident today. To understand that, we'll need to look deeper.

Before doing that, though, let's examine some of the consequences of genetic testing, One near-term result will be a dramatic shift towards preventive medicine. As we learn about our individual genetic vulnerabilities, we will seek ways of mitigating our risks. This will not only mean nutritional and lifestyle interventions, which are notoriously hard to adhere to, but long-term pharmaceutical interventions. This shift will be a huge scientific, clinical, and political challenge. Given that there is only one genetic counselor for every 150,000 people in the U.S., how are we going to understand what we need to do? And once we figure that out, how are we going to pay for the interventions? Our healthcare system is woefully unprepared to deal with the elective, preventive measures that will flow from such genomic screening.

Pharmacogenetics, the tailoring of drugs to our individual genetic constitutions and biochemistries, is another important development. It will no doubt bring more and better drugs, but under the current regulatory environment, they won't be cheaper. In the decades ahead, we will likely spend ever more of our Gross Domestic Product on healthcare and ever more of our healthcare dollars on drugs. How much we're willing to pay for this is going to be a contentious political issue. But the increased spending that these developments bring is not a sign of failure but of success, because we care so much about our health.

This shift towards more personalized medicine is going

to be a huge challenge for the U.S. Food and Drug Administration, and the regulatory apparatus that tries to ensure safety and efficacy will almost certainly be inadequate to the task ahead. It is too slow, too risk averse, and too expensive to serve our needs. As there are fewer block busters and more niche drugs, we will not be able to afford the huge expenses of today's broad gold-standard, double-blind clinical trials. Indeed, the problem of establishing efficacy for personalized interventions is far from trivial at any cost, and there are good arguments that the government should largely abandon it and focus squarely on safety.

Our ideas of political correctness will also be challenged by the information that emerges from broad genetic testing. As we uncover the constellations of genes that help shape us, we will find many differences between individuals, and between populations. Today, it is fashionable to state that only 1 in 1,000 of the base pairs in our DNA differ between individuals, and proclaim that this shows that we all are virtually identical. Of course, this is true: we no doubt would look pretty much the same to a lion looking for a meal or an alien from another planet. We all are animals, all mammals, all primates, all human beings. But the 99.9% of similarities are not what we ourselves care about. We care about the differences among us, and our perceptions are fine-tuned to perceive the subtlest such differences when we're looking for a mate or an employee, or making judgments about other people. We'll have to come to grips with the fact that these differences often have biological underpinnings. Some worry that we are not up to this task and fear that the coming knowledge might tear apart the human family. But much evidence suggests that we can deal with our differences and can do so within the egalitarian framework we so value. The acceptance of differences among individuals is far higher today than it has ever been.

Such immediate challenges, however, are not the source of our deep angst about biotechnology. Our real fears come from more distant possibilities that could be far more difficult and divisive. Three basic realms are apparent.

More Distant Possibilities

The first is the reshaping of our biology. That doesn't mean that we are somehow going to develop another pair of arms or a set of gills. Other aspects of our biology are much more important to us. What if we could unravel the processes of aging and learn to retard or even reverse critical aspects of it? This is a key focus of regenerative medicine, and success would affect virtually every aspect of human society, from family relationships, to educational structures, to the passage of wealth and power from one generation to the next, to the shapes of various social institutions. The potential collapse of social security is the least of it. Nor is this pseudo-science. Steven Austadt, for example, a mainstream biogerontologist from the University of Idaho, believes that the first person to live to be 150 years old is already alive.

People often tell me they don't think it would be a good idea if we were able to extend the human life span. They are worried that there are too many people on the planet already, that there would be environmental consequences, that life would lose its meaning, that extending one's life would be selfish, or even that all those extra years would be boring. Then they often whisper, "but put me on the list."

Extension to human vitality and life span would be widely embraced if they arrived. William Butler Yeats captured our yearning for immortality eloquently in his poem "Sailing to Byzantium" when he wrote, "Consume my heart away; sick with desire and fastened to a dying animal, it knows not what it is. Gather me into the artifice of eternity."

This project to dramatically extend the human health span

is greatly at odds with the present goals of biogerontology, which are not to buy us more life, but essentially to condense our period of morbidity so that we could live healthy lives and then, within a very narrow period, rapidly deteriorate, like a salmon that has spawned. But the full achievement of this goal would be a nightmare, profoundly at odds with our true aspirations. Imagine how unprepared we would be to leave this world if we lived vitally until 80 and then, within a month or so, died. Not only would this be excruciating, it would leave gaping wounds behind, for sickness and decline prepare us and those around us for our departure. I suspect that given the choice, most people would prefer both a longer and a healthier life than either one by itself.

A second area that troubles many people is our increasing reliance on pharmacology not just to heal ourselves, but to manage our emotional states as well. Such "cosmetic psychopharmacology" falls outside of regenerative medicine, but will be increasingly challenging in coming decades. Ritalin, Viagra, Prozac, and other such drugs are only clumsy baby steps. The potential is now emerging to short-circuit the emotional programs that have arisen in our evolutionary history to direct human behavior in ways that further survival and reproduction. It is no accident that we like sugar, that sex feels good, that we are strongly attached to our family, that success can bring a sense of fulfillment.

What would happen if we could take a cocktail of drugs that made us feel contented and fulfilled in whatever we were doing and had no noticeable physiological side effects? Would we be able to resist using it? And if we didn't, who would we be? Why would we do what we do? What would motivate and drive our behavior? Whatever agents we are able or unable to fashion in this arena, it is clear that the lines between medical pharmaceuticals and illicit "recreational" drugs will become ever more obscure in the years ahead.

A third area where biotechnology is challenging us is reproduction. This is not surprising given that the passage of life from one generation to the next is so central to people's perceptions of who they are.

As we understand the constellations of genes that influence our identities, potentials, vulnerabilities, and temperaments, we will want to make choices about the genetic constitutions of our children. Setting aside familiar low-tech methods of mate selection such as arranged marriages and romantic love, and various coming refinements like computer dating, we can anticipate the arrival of three high-tech procedures to consciously choose our children's genes.

The first is reproductive cloning. Media obsession with this is due more to the symbolism of such an alteration of human reproduction than to any plausible consequences of the occurrence. After all, here is a technology that does not yet exist for humans—indeed, no one has yet used nuclear transfer to create viable embryos even in nonhuman primates—and yet enormous global attention has greeted the flimsiest of claims about the procedure. Even more importantly, fear that someone, somewhere, might clone a child has generated serious attempts to criminalize basic biomedical research in regenerative medicine, endangering the hopes of real people with real diseases and real suffering.

Part of this fear no doubt comes from an exaggeration of the consequences of reproductive cloning. Reproductive cloning will likely occur within five to ten years, long before mainstream medical professionals view the procedure as safe, and it will be reported on numerous occasions before it is achieved, as in the case of the Raelian scam in December of 2002. But when reproductive cloning does occur, it is hardly going to be a cataclysmic event. Given the challenges of the procedure, it will necessarily occur on a very small scale, and there will be a long lag before it is seen as safe enough by any but a few driven souls. A few babies may be harmed, of course, but it is hard to see how the creation of a delayed identical twin is going to bring down our civilization. And it is hard to see how fears of cloning could possibly justify stopping embryonic stem cell research, which seems to hold such promise. Our real concern should not be that some fringe group will attempt to clone a child somewhere, but that fears about this will be purposely fanned by those who wish to control the future of reproductive medicine and biotechnology in general.

Another means of selecting our kids' genetic constitutions is embryo screening or "preimplantation genetic diagnosis (PGD)." In this already common procedure, a single cell is removed from a six- or eight-celled embryo in vitro and a genetic test preformed on it. The results then are used to decide whether to implant that embryo or discard it in favor of another. PGD has been in use for ten years to avoid serious diseases such as cystic fibrosis. Such screening procedures are far more important than cloning, and will come into widespread use once in vitro fertilization progresses to the point where it is feasible to freeze immature oocytes. Then, a woman could have a simple ovarian biopsy to collect thousands of eggs, freeze them, and bank them, so that years later, she could thaw and mature them in vitro and fertilize them with her partner's sperm. If she believes in letting nature take its course, she might choose one of the resultant embryos at random for implantation, but I suspect it will be far more common to first screen the banked embryos for genetic diseases or even temperament and personality traits. Such technology will greatly impact human reproduction as it spreads from the infertile who are already using IVF, to the affluent who are worried about genetic disease, to the population at large, and as the technology morphs from a simple screen for disease towards screenings for lesser vulnerabilities like severe depression and then for traits of personality and temperament. Moreover the technology will likely become so potent that the biggest pressure may be to make it available to everyone, rather than to ban it.

The ultimate means of changing human reproduction, of course, is through direct germline intervention—direct alteration of the genetics of the first cell of the human embryo. This would be the beginning of conscious human design. I examine this at length in my book, *Redesigning Humans*, and won't discuss it further here, other than to say that technologies like human artificial chromosomes, which it will require, are not nearly as far away as most people imagine. Direct germline interventions will emerge quite naturally from the sophisticated screening technologies that become feasible in the decades ahead.

Moving Forward

In the manipulation of human embryos for purposes of screening and enhancement, not everything that can be done should or will be done. But international polls show that 25% to 80% of parents in every country examined say that they would enhance the physical or mental capacities of their children through these technologies if they could do so safely. Given that significant numbers of people everywhere regard such technology as beneficial, that it will be feasible in thousands of laboratories throughout the world, and that its use will be so difficult to monitor, the question is not whether such technology will arrive, but when and where it will, and what it will look like.

Humanity will go down this path for two reasons. The first is that there is no need to pursue the technology directly. It will emerge from the basic biomedical research now underway at our most prestigious institutions.

The second reason we will proceed is that we're human and have always used technology to try to enhance our lives. We plant. We build. We mine. We hunt. We dam. And for as long as we have been able, we've altered ourselves as well. We cut our hair, straighten our teeth, pierce our bodies, tattoo our skin, fix our noses. We use drugs to ease pain, lose weight, stay awake, go to sleep, or just get high. The idea that we will long forego better and more potent ways of modifying ourselves is every bit as much of a denial of what the past tells us about who we are, as to imagine that we would use these technologies without fretting about them.

Worrying about these new technologies has considerable survival value, but efforts to ban them will not succeed. And if we draw regulatory guidelines that are too restrictive, it will merely shift research elsewhere, drive it from view, and reserve the technologies themselves for the wealthy, who are in the best position to circumvent restrictions. Such policies thus would mean both relinquishing control over these new technologies by handing them off to others, and delaying whatever benefits emerge from them.

As with so many other extensions of human power, no matter how much debate occurs, there will be no consensus about changing human reproduction, expanding human potential, extending human longevity, using drugs to alter human emotion, or simply screening human embryos. Each will remain contentious because it touches us and our vision of the human future too deeply. Our responses are too driven by our history, religion, culture, philosophy, and politics.

A bioethics advisory commission may pretend that moratoria and additional debate will allow us to reach consensus, but this is delusion. The two poles of the debate are too far apart. To some, changing human reproduction is an invasion of the inhuman. It violates their views of what is most sacred. To others, such advances represent the flowering of our humanity, a chance to transcend aspects of our biology in ways other generations could only dream of.

Weighing the possibilities of regenerative medicine is easy

for those in these extreme camps. Christian fundamentalists will oppose them with all their fiber, and Transhumanists will embrace them wholeheartedly, pushing to have regenerative technologies introduced as rapidly as possible. The people most conflicted by the prospects of changing human reproduction are those who are involved in the work of bringing it about and are concerned about some of its applications.

Ian Wilmut, for example, who cloned Dolly and has probably done more than anyone to bring about mammalian cloning, is deeply troubled by the possibility of its application to humans. It is hard to imagine, though, that a human will not one day be cloned. These technologies will arrive. They will help many people, and they no doubt will cause injuries as well. Such is generally the case with new technology. Our challenge will not be how we deal with cloning, genetic screening, or any specific technology, but whether we have the courage to continue to face the possibilities of the future and accept the risks inherent in such exploration, or whether we pull back in fear and relinquish this effort to other braver souls in other regions of the world.

One of the most dangerous paths would be that represented by the precautionary principle, which posits that we should be certain that technology is absolutely safe before we attempt to use it. This recipe for stasis is not an option, but were it possible, it would deny us the benefits of new developments simply because their development entailed risk. None of the progress of recent centuries could have taken place under a regime embracing such a principle.

The political choices we make today will be very important, not because they will determine whether these new technologies arrive, which is not in question, but because they will determine our ability to influence the development of these technologies. Germany has banned work on embryos and now exerts no influence in that realm.

Our Fears

Let's return to the question we began with: What are we really afraid of? Critics are not particularly concerned that these technologies—including those of regenerative medicine—will fail or that someone somewhere will be injured by them. That would be a relief in many ways, because then the technologies would fade away or at least be greatly slowed.

The real worry is that these technologies will succeed so gloriously that they will create a host of seductive possibilities we can't resist. What strikes fear into the hearts of critics is that we as individuals and as a society will see so much benefit in these technologies that we will embrace them. Then, critics will have to face their real fears:

Their first fear is of ourselves. They worry that we (or rather, someone else) will abuse these technologies. But here, I take comfort from the protections that come from liberal governments and open markets. Together they act against both the grand social engineering projects that totalitarian regimes sometimes inflict on their citizenry and the many bizarre dangers that often populate bioethical debate. The former violate our guarantees of personal freedoms and the latter have too few potential users to inspire commercial development that figures significantly in our future.

Society has navigated technological change before and will muddle through once again. These new biomedical technologies are not like nuclear weapons, where one mistake can vaporize millions of innocents. In truth, we need to make mistakes and learn from them to gain the wisdom to handle these new possibilities wisely. Most people will show common sense when it comes to making choices about these things in their own lives—far more, perhaps, than the government committees whose members do not suffer the consequences of the mistakes they make. But progress in these realms will not be smooth and people will get hurt. That is the way both biological and technological evolution work.

Another fear is the philosophical implications of these coming possibilities. People worry that they may change our sense of who we are. This is probably the least valid basis for making public policy about new biotechnology, because our sense of what it means to be human will so obviously be changed by a broad range of technological advance in the years ahead. If we could stop biotechnology dead in its tracks, other technologies would still reshape our society. When we contemplate the future, we should remember that our great grandparents would not be comfortable in the world of today, and yet most of us would not want to live in their time. So it will be with our future. We may not like it, but our descendants will, and they will likely look back at the present era as primitive and uninviting.

A third fear is that we will be forced to make difficult choices as these new genetic and reproductive technologies arise. But such choices are part of growing up. Humanity is leaving its childhood and moving into its adolescence as our powers infuse into realms that were hitherto beyond our reach.

As we learn to use regenerative medicine to extend our vitality and stave off death, we will need to decide more often when to forgo such technology and let death come. And yet, we are not good at making such decisions even with today's limited technology. When people talk about extending human life span, they frequently begin to worry about immortality. But a chasm lies between doubling our life spans and achieving immortality.

The beginning of life will present us with tough decisions too. Being able to screen embryos means we will have to make judgments about our potential future children and pick one over another, something we are very uncomfortable with.

The biggest fear, though, is that we will have to relinquish control, or rather, acknowledge that we don't really have control over humanity's future. The long-term consequences of the self-directed evolution now underway are not something we can plan, because our course hinges too much on the character of future technologies we cannot yet glimpse and on the values of future humans we cannot yet hope to understand.

It is ironic that accepting the technological path I've described is a deeply spiritual choice. Embarking on this voyage to we-know-not-where is a far greater act of faith than trying to barricade against it by insisting that we should not play God. And, of course, we already do play God in so many ways, from using antibiotics to flying airplanes.

I have mentioned the largest fears of those who would hold back coming advances in regenerative medicine and other realms of biotechnology. But in my view, the dangers from these possibilities pale beside the risk that we will succumb to our fears and pull back from these new technologies, delaying beneficial might-have-beens. Think of the millions of unperceived injuries that would have resulted from policies that delayed the arrival of the polio vaccine for a decade.

We have entered a new millennium, and long before the next, future humans will look back at our era. They may see it as a challenging, difficult, turbulent time, which of course it is, but I think they will primarily see it as the unique, extraordinary, glorious instant when the very foundations of their lives—better and longer lives with better health and greater choices—were laid down. It is a remarkable privilege to be here to observe this critical transition in the history of life. But of course, we are more than observers; we are also the architects of a broad set of changes that will soon reshape human life and our understanding of it. In my view, we should be proud of this. But don't think for a moment that these shifts will not challenge us, because we—like everyone else—are also the objects of these changes.

In 430 BC, the Athenian historian Thucydides seemed to have foreseen our challenge and our calling when he wrote, "The bravest are surely those who have the clearest vision of what is before them, glory and danger alike, and yet notwithstanding go out and meet it."

Cloning Babies Is Not Inherently Immoral

Ronald Bailey

Human cloning hit the headlines when Clonaid, a company associated with the Raëlian UFO cult, announced the dubious birth of a baby girl clone named "Eve" at the end of 2002. It turned out to be hoax, but some day it will not be. After all, South Korean scientists announced in February 2004 that they had succeeded in coaxing 30 cloned human embryos to develop to the blastocyst stage consisting of about 100 cells. The clones were created by means of somatic cell nuclear transfer (SCNT) in which the genes from donated human eggs are removed and then adult cells with all their genes are merged with the enucleated eggs. So the blastocysts are clones of the adults that donated the genetic material. They then removed the inner cell mass from 20 of the embryos and were able to establish one colony of ES cells. The cells from the inner cell mass are pluripotent, that is, they can differentiate into all the diverse types of tissues that form the human body.

Until this past February, no scientists have been able to grow cloned human cells to the blastocyst stage, much less create a colony of cloned ES cells. ES cells are highly desirable for transplants since they are nearly genetically identical (except of mitochondrial DNA) with the cells taken from the donor. This means that such cloned cells would be perfect transplants because they would be unlikely to be rejected by a patient's immune system.

However, instead of using the blastocysts to produce stem cells, the Korean researchers could have tried implanting them into a woman's uterus to produce a cloned child.

Cloning people became a realistic prospect following the birth of the cloned Scottish sheep Dolly in 1997. Eve was allegedly cloned using techniques similar to those that produced Dolly. In Dolly's case scientists combined the nucleus taken from a sheep's mammary cell with a sheep's egg, from which they had removed its nucleus. It was immediately apparent that if researchers can clone a sheep, surely they can clone a human being, too.

But let us set aside the therapeutic advantages offered by the creation of stem cells by means of cloning to concentrate on the question of what exactly, if anything, is wrong with cloning to produce human babies? After all, both Presidents Bill Clinton and George Bush have called for a ban on human cloning and the U.S. House of Representatives has twice voted to criminalize all human cloning, both therapeutic and reproductive, by fining perpetrators $1 million and throwing them in prison for up to ten years.

Perennial Luddite Jeremy Rifkin, head of the Foundation on Economic Trends, grandly pronounced that cloning "throws every convention, every historical tradition, up for grabs. "At the putative opposite end of the political spectrum, conservative columnist George Will chimed in: "What if the great given—a human being is a product of the union of a man and woman—is no longer a given?"

In addition to these pundits and politicians, a whole raft of bioethicists have declared that they, too, oppose human cloning. Daniel Callahan of the Hastings Center said flat out: "The message must be simple and decisive: The human species doesn't need cloning." George Annas of Boston University agreed: "Most people who have thought about this believe it is not a reasonable use and should not be allowed.... This is not a case of scientific freedom vs. the regulators." In fact, Annas has been leading a campaign to get the United Nations to adopt a treaty outlawing reproductive cloning.

In 2002, President Bush's Council on Bioethics issued a report that recommended that all human reproductive cloning be completely banned and urged that a four-year moratorium on therapeutic cloning research aimed at producing stem cells also be adopted.

Given all of the brouhaha, you'd think it was crystal clear why cloning humans is unethical. But what exactly is wrong with it? Which ethical principle does cloning violate? Stealing? Lying? Coveting? Murdering? What? Most of the arguments against cloning amount to little more than a reformulation of the old familiar refrain of Luddites everywhere: "If God had meant for man to fly, he would have given us wings. And if God had meant for man to clone, he would have given us spores." Ethical reasoning requires more than that.

What would a clone be? Well, he or she would be a complete human being who happens to share the same genes with another person. Today, we call such people identical twins. To my knowledge no one has argued that twins are immoral. Of course, cloned twins would not be the same age. But it is hard to see why this age difference might present an ethical problem—or give clones a different moral status.

You should treat all clones like you would treat all monozygous [identical] twins or triplets," concludes Dr. H. Tristam Engelhardt, a physician and bioethicist at Rice University. "That's it." It would be unethical to treat a human clone as anything other than a human being. If this principle is observed, he argues, all the other "ethical" problems for a secular society essentially disappear. John Fletcher, a professor of biomedical ethics in the medical school at the University of Virginia, agrees: "I don't believe that there is any intrinsic reason why cloning should not be done."

Let's take a look at a few of the scenarios that opponents of human cloning have sketched out. Some argue that clones would undermine the uniqueness of each human being. "Can individuality, identity and dignity be severed from genetic distinctiveness, and from belief in a person's open future?" asks George Will.

Will and others have apparently fallen under the sway of what Fletcher calls "genetic essentialism." Fletcher says polls indicate that some 30 percent to 40 percent of Americans are genetic essentialists, who believe that genes almost completely determine who a person is. At its worst, some bereaved and misinformed people apparently have the deep misconception that cloning can bring back or replace a dead child or other loved one.

However, considering again the case of the natural clones called twins helps us think clearly about what clones produced in vitro would be like. Twins are clearly distinct individuals with different points of view because twins have two different bodies and two different brains. A person who is a clone would live in a very different world from that of his genetic predecessor. With greatly divergent experiences, their brains would be wired differently. After all, even twins who grow up together are separate people—distinct individuals with different personalities and certainly no lack of Will's "individuality, identity and dignity." Individuality does not reside in our genes, but in our brains and bodies.

In addition, a clone that grew from one person's DNA inserted in another person's host egg would pick up "maternal factors" from the proteins in that egg, altering its development. Physiological differences between the womb of the original and host mothers could also affect the clone's development. In no sense, therefore, would or could a clone be a "carbon copy" of his or her predecessor.

What about a rich jerk who is so narcissistic that he wants to clone himself so that he can give all his wealth to himself? First, he will fail. His clone is simply not the same person that he is. The clone may be a jerk too, but he will be his own individual jerk. Nor is Jerk Sr.'s action unprecedented. Today, rich people, and regular people too, make an effort to pass along some wealth to their children when they die. People will their estates to their children not only because they are connected by bonds of love but also because they have genetic ties. The principle is no different for clones. But the considerable emotional and financial investment that the parents of cloned children will be making indicates that these children will be very much wanted and treasured by their families.

Some cloning opponents worry about a gory scenario in which clones would be created to provide spare parts, such as organs that would not be rejected by the predecessor's immune system. "The creation of a human being should not be for spare parts or as a replacement," declared Senator Christopher Bond (R-MO). I agree. The simple response to this scenario is: Clones are people. You must treat them like people. We don't forcibly take organs from one twin and give them to the other. Why would we do that in the case of clones?

But what about cloning exceptional human beings? George Will put it this way: "Suppose a cloned Michael Jordan, age 8, preferred violin to basketball? Is it imaginable? If so, would it be tolerable to the cloner?" Yes, it is imaginable, and the cloner would just have to put up with violin recitals. Kids are not commercial property—slavery was abolished some time ago. We all know about Little League fathers and stage mothers who push their kids, but given the stubborn nature of individuals, those parents rarely manage to make kids stick forever to something they hate. A ban on cloning wouldn't abolish pushy parents.

One putatively scientific argument against cloning has been raised. As a National Public Radio commentator who opposes cloning quipped, "Diversity isn't just politically correct, it's good science." Sexual reproduction seems to have evolved for the purpose of staying ahead of ever-mutating pathogens in a continuing arms race. Novel combinations of genes created through sexual reproduction help immune systems devise defenses against rapidly evolving germs, viruses, and parasites. The argument against cloning says that if enough human beings were cloned, pathogens would likely adapt and begin to get the upper hand, causing widespread disease. The analogy often cited is what happens when a lot of farmers all adopt the same corn hybrid. If the hybrid is highly susceptible to a particular bug, then the crop fails.

That warning may have some validity for cloned livestock, which may well have to live in environments protected from infectious disease. But it is unlikely that there will be millions of clones of one person. Genomic diversity would still be the rule for humanity. There might be more identical twins, triplets, etc., but unless there are millions of clones of one person, raging epidemics sweeping through hordes of human beings with identical genomes seem very unlikely.

But even if someday millions of clones of one person existed, who is to say that novel technologies wouldn't by then be able to control human pathogens? After all, it wasn't genetic diversity that caused typhoid, typhus, polio, or measles to all but disappear in the United States. It was modern sanitation and modern medicine.

But should attempts to produce human babies go forward now? After all, renegade fertility specialists, Severino Antinori and Panayiotis Zavos have been periodically claiming over the past 3 years that they are trying to produce a baby by means of cloning. However, if they or other would-be human cloners succeeded now they would likely bring a very sick and defective baby into the world. Why? Because one of the chief problems in cloning mammals so far has been its inefficiency—it takes a lot of failed embryos to produce one healthy cloned animal. Right now, only 2 percent to 4 percent of mammalian clones are long-term survivors, according to Massachusetts Institute of Technology Biology Professor Rudolf Jaenisch. Why such a low survival rate? Jaenisch points out that many cloned mammal fetuses, e.g., pigs, calves, and sheep develop severe abnormalities. For example, some animal clones produce larger than normal placentas, others are born twice as big as normal, and some are born with deadly anatomical flaws, like enlarged hearts or defective kidneys.

Jaenisch suspects that the abnormalities result from parental genomic imprinting. To make a long story short, the problem with mammalian cloning may be that the genes from nuclei from mature cells may have lost their proper imprinting due to aging. So when these mature nuclei are inserted into enucleated eggs to produce embryos, their imprinting is wrong. Since there is currently no way to restore it, either paternal or maternal genes affecting fetal growth may end up being dominant creating the developmental imbalances seen in cloned animals. There is no test now available for checking on whether genes are properly imprinted or not, so bringing a healthy clone to term is largely a matter of chance.

Still, much progress is being made in the cloning of animals. Among the species successfully cloned are cows, pigs, sheep, goats, mice, rats and cats. Australian biologist Ian Lewis at Monash University has cloned award-winning Holstein milk cows. Cloning such champions could boost milk production. Other groups have cloned award-winning beef cattle. The Kelmscott Rare Breeds Foundation in Maine has used cloning to rescue a rare breed of pig, Gloucester Old Spots, from extinction. Spanish researchers are planning to use cells from the recently extinct Bucardo mountain goat to bring the species back to life. David Ayares of PPL Therapeutics located in Virginia has cloned genetically engineered pigs whose organs have been "humanized" by removing the surface sugars that trigger immune rejection. This is a step toward producing animal organs suitable for transplantation into people.

Research in animal cloning perfects cloning techniques in general. As scientists understand more about what so often goes wrong in cloning, they are learning how to prevent such errors. Safe human cloning, like safe human in vitro fertilization, will arise from advances in animal research. British fertility researchers Robert Edwards and Patrick Steptoe spent decades refining in vitro fertilization techniques in a variety of animal species. Only after they were certain those techniques were safe did they attempt to use them in people. The result of their research was the birth of Louise Joy Brown in 1978. Some estimate that nearly 1 million healthy babies have been born worldwide using IVF over the past quarter century.

So, instead of trying to create human babies through cloning now, most researchers agree it would be far better for research to continue on cloning other species until scientists understand why the abnormalities occur and can reliably prevent them. A good benchmark for deciding when to proceed with human reproductive cloning would be when researchers are reasonably sure that clones would suffer no more likelihood of birth defects (about 2%) than do children produced by sexual reproduction. It's too early now, but few doubt that the technical barriers to safely cloning human beings will fall in the next decade or so.

Once the public understands the limitations of reproductive cloning – for example, one can't bring back the dead or create a newer, younger version of oneself – human cloning will likely be used mostly by infertile and gay couples who have no other way to bear biologically related children. Far from seeing cloning as immoral, University of Texas bioethicist John Robertson argues that parents have a fundamental human right to use biotechnology to help them bring healthy children into the world. It is likely that more and more people will come to understand that cloning to produce a healthy child is not inherently immoral. After all, a human clone is not a monster, but simply a person who shares his or her genome with another person.

A Model for Regulating Cloning

Glenn McGee and Ian Wilmut

Regrettably, there may be individuals on earth who would find the prospect of participation in clinical trials of human reproduction through somatic cell nuclear transfer acceptable or even appealing.

Odd, imaginative, and unlikely examples have been proffered. In one, a woman whose husband is killed seeks to clone another of her already born young children to have another child with the husband.[1] In another a young couple, whose genes code for the lethal Tay–Sachs disease, requests nuclear transfer so that germ-line gene therapy can be conducted on the embryo to remove the condition. In still a third, the extremely rare woman with disease in the mitochondria of her cells might request cloning with donor egg in order to avoid passing on her lethal disorder through her own egg. These odd, hard cases miss the more general context within which any request for human cloning might originate.

Those who encounter fertility problems not easily remediable by therapy to the reproductive organs can walk several paths. Each has risks and benefits for the individuals, couples, and families involved. Some choose to play supportive or parenting roles in their extended families, or marry into families with children. Some enroll in adoption programs. Others turn to clinical medicine, where a wide and expanding range of techniques is aimed at providing a pregnancy and eventually a child. It is the hope of those who choose this last set of options that the resultant children will share some or many hereditary traits with his or her parents.

For patients with some kinds of infertility affecting the gametes, the use of donated sperm or eggs can greatly improve the likelihood of successful fertilization and pregnancy. Choosing to utilize donated gametes or embryos carries benefits and hazards. The benefit is a child that shares some of its parents' genes and in many cases a fetus that can perhaps be carried by its eventual mother. However, there are important long-term risks associated with donor gametes. Even the best screening of donors cannot rule out the presence of hereditary risks for disease that are hidden in the donor's DNA, risks the donor may not know. The child may want or even need to know about the medical and social history of his or her "donor parent." The existence of an additional donor-parent may present long term problems for the child and family. This issue may be particularly acute when viewed against the backdrop of the goals of in vitro fertilization, namely to preserve a strong genetic link in the nuclear family.

Individuals and families may at some point present their gynecologists, urologists, and infertility specialists with requests for somatic cell nuclear transfer. The challenge of such requests must be set in broad international and interdisciplinary context. The international debate about safety in clinical cloning research is a significant first step toward debate among scientists, clinicians, clergy, and the public about the panoply of new discoveries in reproductive science and medicine. Significant questions are coming into focus: Who is best suited to ensure the safety of children born through new reproductive technologies, and how ought they to make decisions? What relationships should exist between parties who participate in new reproductive technologies and how ought such relationships to be consecrated? We argue that answers to these and other questions should be framed not by broad governmental paternalism in science and medicine, but instead on the model of progressive, regional social oversight aimed at protecting the long-term interests of children. We call this the adoption model, and in this essay describe and defend its application to human cloning law and ethics.

It is patent that human cloning should not proceed to the clinical research stage. A moratorium on clinical trials of human cloning is warranted on safety grounds, as there is no pathway from animal to pre-clinical to clinical human experimentation that would not involve significant risks to human children.[2] As we have noted elsewhere, it is doubtful even in the long term that an individual or couple will present a rationale for the use of human cloning technologies that is compelling when balanced against the risks.[3] In this essay a scientist and an ethicist argue that social restrictions on human cloning can best be justified and implemented on the model of law and policy about adoption.

Regulation and debate about human reproduction may be modeled on three different emphases. We will call these models the **reproductive freedom model,** the **pediatric model,** and the **adoption model**.

This century has seen the birth of an entirely new kind of jurisprudence about sexuality and reproduction. Fueled by scientific developments like birth control and in vitro fertilization, and against the backdrop of international civil rights reform, courts in the 20th century framed a new dimension of the freedom of expression: the right to choose one's progeny. The right to make one's own decisions about reproduction has several strata. A right against government interference in reproduction is most clearly codified in American case law about discontinuation of pregnancy. *Roe v. Wade* and *Casey* carve out a right to reproductive privacy and link pregnancy to other central human expressions of flourishing.

Indeed the central tenet of reproductive freedom is the fairly obvious fact that the reproductive life is central to self-identity, flourishing, and free expression more generally, for individuals and for families. While marriage is highly regulated, as are numerous sexual practices, no license is required for childbearing.

Because families and individuals have such broad freedoms in making children, advocates of reproductive freedom have maintained that it would be inappropriate, even discriminatory, to apply special restrictions to those who are infertile. Why ought the infertile person to be forced to undergo special screening prior to pregnancy when individuals whose reproductive capacity is intact can initiate pregnancy in the most unorthodox ways imaginable without fear of social scrutiny?

The argument against state interference in reproduction is a negative freedom.[4] Arguments for positive freedoms in reproduction, for entitlement to reproductive services, proceed apace. The standard of care for treatment of infertility is not obvious. While the clinical dimensions and etiology of a patient's infertility may be apparent, if the patient's underlying pathology cannot be treated (e.g., the testes repaired), it becomes unclear what the "cure" will be. Is the patient who has two children through in vitro fertilization cured? Which techniques, with what ends, should augment or substitute for reproductive capacity? Any answer to these questions will be textured by subjective considerations such as patients' ability to pay for services, the allocation of government resources to infertility research and treatment, and the technological limits of existing treatments.

Those who advocate the primary role of reproductive freedom in the human cloning debate point to the importance of allowing individuals and families to think for themselves about having children.[5] If the state allows couples to have children in squalor or single parent families, how can it reasonably proscribe human cloning as either unsafe or irresponsible? Some American scholars go so far as to argue that U.S. restrictions on human cloning would violate the Americans with Disabilities Act, a law prohibiting discrimination against, in this case, the infertile.

On the opposite side of the human cloning fence are those who argue that human cloning would in some way harm children, and should be prevented in the interest of safety. The argument is made in the spirit of what we call the pediatric model, which emphasizes not the rights of procreators but the responsibility to care for those created.

In the 20th century, healthcare for very young and very vulnerable children has become such a high priority as to rate inclusion in public health policy around the world. To see the enormous changes in the meaning and status of children one need only turn to the copious literature on the creation of childhood as an institution, which takes place during this century. Where one hundred years ago children made up a significant segment of the workforce, and infant mortality was a staggering 30–40% in some nations, today many parents can expect that their children will have access to comprehensive medical and educational resources. The identification of children's needs begins early, and among the very best selling books in the world are guides to pregnancy and early childhood. Incredible amounts are spent on neonatal intensive care and high-risk obstetrical care, as the most vulnerable infants imaginable are kept alive even after extremely premature birth.

In ordinary terms, far from the tertiary care hospital, the pediatric metaphor is felt in public practice and policy about pregnancy and childhood. Parents in many nations come to know their child not as a child but as a fetus, with interests even early on in pregnancy. This can be mundane—an unnecessary ultrasound examination recorded on videotape to allow a mom to show the entire family her 8-week fetus. The presence of the fetus as an organism with interests has also presented extraordinary new problems. In *In re A.C.* the U.S. Court ruled that the right to discontinue pregnancy does not include a concomitant right to willfully harm the fetus. In hospitals this can mean that women are assigned to social workers early in their pregnancy on the basis of their drug habits or other problems. They can choose to discontinue pregnancy without interference, but if they elect to bear the child in an environment that is dangerous to the child, steps may be taken in advance toward removing the future child at birth from the care of the mother.

The paradigm for such remarkable steps is the broad social consensus about the need to protect young and vulnerable children from dangers against which they cannot protect themselves. Parents who abuse or neglect children, who refuse to educate children, or who will not provide their children with medical care (like vaccinations) can lose their parental roles. In this important sense, parenthood has always been both a responsibility and a privilege, rather than right, from the view of the pediatric model.

Arguments of the National Bioethics Advisory Commission and others that clinical human cloning should be prohibited have relied on the pediatric model. Two kinds of claims have been made-first that cloning would be physiologically unsafe for any human clone, and second that cloning would deprive a child of its identity or in other ways rob it of freedom. In the first case, it is clear that the claim is pediatric in character. Just as parents are forbidden from intentionally exposing their children to great, preventable risks, it is argued that parents ought not to expose future children to the sorts of hazards experienced by the first offspring in animal human cloning experiments. This duty obtains especially in early trials, when parents could have little or no confidence that their actions would be safe for resultant offspring.

In the second case, the argument of Dena Davis and others that children have "a right to an open future"[6] is also based on a social commitment to ensure that those who make children participate in certain pursuits taken by the community to be essential for the development of children. It is required in many nations that children be educated. The failure to provide children with clothing or a home safe from extreme violence is punishable throughout the world. In this regard it is feared that cloning might put children in an untenable family relationship or rob them of skills necessary for flourishing. The young clone might grow up with his or her progenitor as a "living" genetic test, knowing early on what is in store for his or her own future.

The litmus test for human cloning, from the pediatric perspective, is the interest of the clone. If it can be argued that the human child born through a new reproductive technology will be significantly imperiled in a preventable way, those who argue for the interests of the clone will hold that the procedure was unwarranted.

While arguments in the pediatric model seem very valuable to us, neither the pediatric nor reproductive rights model speaks to the question of how to regulate or debate human reproductive technology. Thus while we may agree with advocates of reproductive liberty that parents ought to have wide latitude in their sexual and reproductive choices, it is unclear how one would recognize any compelling interest that merits restricting that latitude. And while we may agree that the cloning of humans does not take sufficient account of the interests of the clone, it is unclear how to prevent similar tragedies from taking place in low-tech parenthood, or how to regulate new reproductive technologies so that disaster is averted.

One significant impediment to dialogue between those who argue for reproductive rights and those who argue for the interests of the child is the dilemma described by Derek Parfit. Because there is no child actually born at the time of the request for clinical human cloning, it at first seems odd to ask whether "the child" is well served by that procedure. One can only with difficulty protect the interests or rights of an organism that does not yet exist. Some even maintain that the debate about the interests of future generations must always be framed in terms of whether or not the future child would be better off never having existed. It is an apparent dilemma. However, there is one area of social policy where the gap between reproductive rights and the interests of children has been nicely bridged, resulting in significant consensus about how to protect children from dangerous situations.

Children have been adopted for thousands of years, and relationships between adoptive families and children have taken many forms and been articulated in many ways. The enormous institutional wisdom accumulated in what we call the adoption model can be very important for bridging the gap between reproductive liberty and pediatrics. The adoption model can move the debate about cloning and new reproductive technologies from its present, highly politicized rancor into a more constructive arena in which interdisciplinary and bipartisan consensus may be possible.

Parents who seek to adopt children are required, in virtually every nation, to seek prior approval from a regional authority or court. In many nations applicants are required to undergo psychological testing, home visits or other prescreening. In most cases these pre-screens take place before a particular child has been identified for adoption; in many cases the pre-screen is independent and antecedent to the identification of a pregnant birth parent.

From the reproductive rights model, it might seem odd that such gross oversight is permitted. After all, fertile parents are not pre-screened before the state permits pregnancy. It could be argued that the screening of applicants for adoption is a manifest invasion of reproductive privacy and an incursion on the rights of parents to reproduce in the manner they desire. We might very well have converted adoption to the model of reproductive rights, following for example the U.S. precedent of leaving surrogacy and egg and sperm donation to the marketplace. Why, when we tolerate a virtual free market in all donor-assisted reproduction, with no pre-screen or judicial oversight, do we insist that adoption be so closely monitored?

The answer is that adoption, in many respects, embodies the best features of both the reproductive rights and pediatric models. Adoption law is framed out of a recognition that the adoption of a child is an unusual way to enter into a family, devoid of pregnancy and birth and textured by its own social and moral features. The adoption process cannot replace these elements of gestation and preparation for childbirth. However, in an important sense it gives communal imprimatur to the creation of a family, drawing on other social rituals for sealing a permanent and loving commitment (i.e. marriage).

The adoptive parents are not screened in search of perfect parents, only with the aim of determining whether or not this particular set of parents can provide some bare minimum features of parenthood that have been historically important in the adoption setting. In this respect the adoption judge is much like the divorce court judge. When parents split up, a judge is in the unusual position of determining what sort of family is best for a particular child given some set of exigencies. What appears to us to be Solomon's wisdom embodied by such judges is actually the product of long-term study of human families in a particular communal context. While their decisions are imperfect, the ethical responsibility of the judge is identified with the representation that the judge makes for the community and for the laws of the state or nation as they apply to adoption.

The adoption judge or magistrate is in an important sense a community historian for the dimensions of family, tracking some of the important features of the community so that they can be accounted for in matching parents and children. Parents who are not judged to be good candidates for adoption may plead their case, but are finally at the mercy of the community leaders.

The adoption model for human family making is predicated on several simple and profound assumptions. First, where unorthodox parenting arrangements (as in adoption or divorce) pose special challenges, the responsibility of the community to provide counsel and oversight is compelling. Second, where arrangements for parenting have not worked or are likely to present special problems, the court and community ought to be empowered to enact short- or long-term restrictions on certain kinds of family-making. Just as regional governments decide how marriage will work, who may inherit, and what kinds of schools provide sufficient education, the family courts have a quite proper jurisdiction in prohibiting certain kinds of family relationships (e.g., incest, cloning, and polygamy). Third, the formation of a family is both a deeply personal and profoundly social act. The interests of children who are adopted or made through new reproductive technologies are best served when a spirit of openness and honesty about the meaning of the process is demonstrated.

It is already clear that we join dozens of other ethicists and scientists in favoring a short-term ban on clinical human cloning. The purpose of this paper however is to argue for a way in which human cloning restrictions might take shape. In their haste to pass legislation, many have settled for a simple, totalitarian approach to a cloning ban. We propose a more democratic, consensus-oriented model that entrusts the community to develop and enforce rules for the protection of children. Even those who disagree with our model will need to argue not only for restrictions on cloning but for the most sensible and careful way to frame such law.

The adoption model can be easily adapted to a variety of reproductive technologies.[7] Our purpose here is not to argue for specific policies. We have explicated the framing of the debate both about human cloning and reproductive technology more generally, holding that while new reproductive technology can be discussed in terms of either reproductive rights or pediatric interests, the two kinds of arguments can seem incommensurable. Adoption as a model integrates both the importance of the rights of parents and the importance of the interests of children, even those children who have not yet been born or even conceived. Where unorthodox parenting and family making is concerned, the community should draw on much richer metaphors than simple analysis of rights. The conflict between reproductive rights and interests of the child is deceptively simple, reflecting more general debates in society about the role of the state in personal life.

By contrast, the making of children is as complex and confusing an area as exists in human inquiry and human life. In adoption, somehow consensus has been reached that children of new and unusual techniques merit special protection, but such protection ought not to be onerous for parents once the parental relationship is consecrated. Moreover, by applying the adoption model to the problem of human cloning, it becomes immediately clear how difficult it would be for any of the test cases described above to meet the high standards for use of such a risky technology. Parents who present with requests that would either excessively stylize children or place them in harm's way ought not to be allowed to proceed. In the short term this will doubtless mean that under an adoption model, sponsored by state or regional governments, cloning ought to be proscribed. At the same time, unlike other larger plans designed to restrict a broad swath of scientific research, the adoption model is a more limited endeavor whose scope is the making of families.

Notes

1 Greg Pence, *Who's Afraid of Human Cloning* (New York 1998).
2 See the recommendations of the National Bioethics Advisory Commission report, included herein. We note too that restrictions on cloning must be crafted carefully so as to ensure freedom in scientific research not intended to produce human children.
3 McGee makes this argument in *The Perfect Baby: A Pragmatic Approach to Genetics* (New York 1997), epilogue; it has been made by Wilmut and others as well.
4 C.f. Arthur Caplan, *Am I My Brother's Keeper* (Indianapolis 1998).
5 See especially Lee Silver, *Remaking Eden: cloning and beyond in a brave new world* (New York 1997) and Philip Kitcher, *The Lives to Come* (New York 1997).
6 Dena Davis, "The Right to an Open Future," *Hastings Center Report*, March–April 1997, pp. 34–40.
7 See, e.g., Glenn McGee and Daniel McGee, "Nuclear Meltdown: Ethics of the Need to Transfer Genes," *Politics and the Life Sciences*, March 1998, pp. 72–76.

The Future Is Later

Chris Mooney

In his recent book *Our Posthuman Future: Consequences of the Biotechnology Revolution*, Francis Fukuyama writes, "Cloning is the opening wedge for a series of new technologies that will ultimately lead to designer babies... If we get used to cloning in the near term, it will be much harder to oppose germ-line engineering for enhancement purposes in the future." For this reason, argues Fukuyama, the current cloning debate amounts to "an important strategic opportunity to establish the possibility of political control over biotechnology." The buck—or, if you prefer, the biotech—must stop here.

By these lights, Fukuyama can't be pleased with the way the cloning issue is faring in the U.S. Senate. As we go to press, it appears that Republican Senator Sam Brownback of Kansas and Democratic Senator Mary Landrieu of Louisiana lack the votes to pass their Bush-backed bill criminalizing both reproductive cloning and the cloning of human embryos for medical research (often called "therapeutic cloning"). Indeed, the Senate may be closer to passing a rival bill that would only ban the use of cloning to produce a full-grown baby. With negotiations collapsed, Majority Leader Tom Daschle has no further plans to bring up the legislation, and a dejected Brownback may carve it up into amendments.

Then turnabout here—and the implicit rebuke to Fukuyama and his ilk—is astonishing. Just last summer, the House of Representatives voted 265 to 162 for a sweeping Brownback-style ban. And since then, as anti-cloning advocates began heaping pressure on the Senate to follow suit, a Fukuyama-esque cloning-as-wedge strategy seemed all pervasive. The "slippery slope" argument has been repeated constantly, for example, in *Weekly Standard* editorials ranting about the danger of a "Brave New World" and even drawing parallels between the head of the biotech firm Advanced Cell Technology and Osama bin Laden. The desire to avert a "posthuman future" also drives the so-called secular case against therapeutic cloning outlined by *Washington Post* columnist Charles Krauthammer (a member of the President's Council on Bioethics, along with Fukuyama), ethicist Leon Kass (the council's chairman), and other neoconservatives.

All of these thinkers support the Brownback–Landrieu bill, and its defeat will be, in large part, their own. That makes it all the more interesting that *none* of them profess to oppose research cloning on the anti-abortion grounds that human embryos—created and then destroyed for their stem cells in the process—are morally equivalent to persons. This is no accident. "There was a real effort to get this off the issue of where life begins," notes Daniel Perry, executive director of the Alliance for Aging Research, which supports research cloning. "Instead, they wanted to shift it to 'the mad scientist versus the people'."

To that end, Fukuyama and the *Standard* have been more inclined to seek alliances with scattered anti-cloning environmentalists and feminists than with the National Right to Life Committee. And they have sought to advance their cause by stoking a primal unease about cloning that springs from our literature (*Brave New World*), myth (*Faust*), and popular culture (*Attack of the Clones*). The idea seems to have been that, due to its resonances, cloning could serve as a Trojan horse for an array of speculative anxieties about the future. As former Clinton administration bioethicist R. Alta Charo puts it, according to this strategy "the cloning debate is about everything but what it's about."

In the Senate, however, the cloning debate turned out to be about exactly what it's about. Largely owing to the education efforts of the umbrella Coalition for the Advancement of Medical Research, legislators who might once have thought of *The Boys from Brazil* when they heard the word "cloning" now think of potential cures for Parkinson's and Alzheimer's. Considering that this is the first full-fledged bioethics battle of the new century—Bush defused the last one with his contrived stem-cell "compromise"—the development is a remarkable one. Fear mongering has been tried and found wanting.

The Senate acted differently than the House for a number of reasons. One is that, as Hastings Center President Thomas Murray puts it, "There's a world of difference in the deliberativeness with which the two chambers have acted in this case." (When the House voted last summer, it did so after just a few hours of discussion.) The pro-therapeutic cloning side also saw some stunning conversions to its position—Orrin Hatch, Strom Thurmond, Nancy Reagan, Gerald Ford—and was bolstered by support from Nobel laureate scientists and celebrities such as Michael J. Fox and Christopher Reeve (who represent an army of patients suffering from serious and often life-threatening degenerative diseases).

The pro-research cloning camp also caught a lucky break with the Catholic Church's pedophilia scandal. Though the Church has traditionally been an 800-pound gorilla on all matters embryo, it has recently lost its claim to the moral high ground. As University of Pennsylvania bioethicist Arthur L. Caplan says, "You try telling somebody you're concerned about embryos these days in the average church, and they're going to be standing up screaming at you, 'Are you concerned about children?'" Though many hardcore Catholics remain active on the issue, the pope has not issued a statement on cloning the way he did on stem-cell research. A bishop's campaign is unthinkable.

Still, the intellectual collapse of the Kass-Fukuyama-Krauthammer secular argument against therapeutic cloning—which was clearly designed to extend the position's appeal beyond the antiabortion crowd—remains the key to the Brownback bill's weak showing. Writers such as Krauthammer (and, for that matter, Republican Senator Bill Frist) repeatedly express support for the president's stem-cell decision but opposition to research cloning. That logic baffles even many anti-abortion intellectuals. As *National Review*'s Ramesh Ponnuru wrote in response to a vast Krauthammer *New Republic* cover story opposing therapeutic cloning: "Krauthammer believes that [b]ecause [an embryo] is not a mere thing, it cannot be created for the sole purpose of using it in a way that destroys it. If it's already been created for some other purpose, though, as the leftover embryos in IVF clinics have been, it can be destroyed. If there's a point of principle that underlies this set of positions, I can't see it."

Indeed, Caplan describes the secular case against research cloning as "incoherent." If Krauthammer and his allies are so concerned about the human manipulation of natural reproductive processes, Caplan suggests, they should not only oppose IVF but "should be lying awake at night terrified about altered soybeans," to say nothing of amniocentesis. In fact, decades ago Kass *did* oppose in vitro fertilization, and for precisely the same sorts of reasons that he now arrays against therapeutic cloning—prominent among them, his argument for the "wisdom of repugnance" (if it seems icky, don't do it).

Given these lines of thought, perhaps it's no wonder that Kass's President's Council on Bioethics has failed to reach any consensus on therapeutic cloning. Nor, consequently, has it has been able to throw real intellectual weight behind the president's and Brownback's position.

The people who *aren't* confused or inconsistent in opposing cloning, though you may disagree with them, are the anti-abortionists. And as the secular anti-cloning argument collapses under its own weight, it becomes increasingly clear that, at its base, the cloning issue boils down to abortion politics by other means. As one therapeutic cloning advocate puts it, "In some offices in the Senate, this whole issue has not even been handled by the health [legislative assistant]." Rather, it was kicked upstairs because of the abortion implications.

This should not come as a surprise to anyone. After all, a Pew Research Center survey in April found that "religious commitment is the most important factor influencing attitudes of opponents of stem cell research." Why cloning opponents would be any different is hard to figure, given that both debates center on embryos.

Moreover, like *The Weekly Standard* crowd, the anti-abortion movement has made a strategic choice to use cloning to its advantage. "To the deep thinkers in the right-to-life movement, stem-cell research and cloning are just one of many issues that they can use to advance opposition to abortion in the long run," notes John C. Green, a political scientist at the University of Akron who specializes in religion and politics.

It's not that the resolution of the cloning issue bears directly on *Roe v. Wade*. But for an increasingly pragmatic and opportunistic anti-abortion movement, the hope is that a legal precedent protecting cloned human embryos would feed into a broader anti-abortion political climate. In this sense, right-to-life opposition to research cloning should be classified alongside the Bush administration's controversial draft regulations to cover embryos and fetuses under the Children's Health Insurance Program and the House's 2001 passage of the Unborn Victims of Violence Act—both of which were widely viewed as indirect attempts to roll back abortion rights.

If Brownback–Landrieu is indeed dead, and if both secular and anti-abortion opposition to therapeutic cloning have indeed been bested—still a big if—where does that leave us? For liberals, there may be an opportunity down the road to push for the federal funding and regulation of embryo research, including therapeutic cloning, as a central government function and concern. "Let me make this modest prediction," says Dartmouth bioethicist Ronald Green, author of *The Human Embryo Research Debates*. "If therapeutic cloning really continues to prove itself as a modality for therapy...then I think you're going to hear voices saying that the [National Institutes of Health] should be overseeing some of this research, and not leaving all the patents, and the opportunities, to the private sector."

That's roughly the way things currently stand in Great Britain. And if it happened here, liberal advocates of research cloning would no longer have to force themselves into a strange alliance with the biotech industry in the interest of protecting science. Once we put the government in control of carefully *regulating* embryo research instead of trying to ban it in the private sector, the United States would also be well positioned to keep pace with other nations in this field. And there's nothing scary—or posthuman—about that future.

Protestant Objections to Human Cloning

Ronald Cole-Turner

Like most other people, Protestant Christians object to human reproductive cloning out of concern for safety and out of fear of the unknown. Beyond this, however, Protestants have additional objections to cloning. This chapter looks at three Protestant objections to cloning and at the theological assumptions on which they are based. The three objections are that cloning confuses basic human relationships, objectifies or dehumanizes the clone, and is a theologically inappropriate response to grief.

It must be asked at the outset what exactly these objections are meant to do. Are they marshaled as arguments to convince Christians, and perhaps others, that reproductive cloning must be permanently banned? Are they meant to convince Christians that they must refuse cloning for themselves even if they tolerate it for others? Or do the objections merely point out concerns that Christians have about how cloning could be misused, and do they function more as guides than prohibitions, as considerations for safe use rather than arguments against all use?

It is probably true that most Protestants believe these objections demand a ban at least for themselves if not for all. But as I will suggest, the objections fail if they are meant to persuade that all reproductive cloning must be banned or even that Protestants must prohibit it categorically for themselves. They succeed, however, in identifying points to consider, a kind of checklist for self-examination and reflection by Protestant clergy and congregations.

Cloning as Confusion

The first objection to cloning is that it confuses human relationships. The point here is not the trivial one, that if for instance I clone myself, my clone is at once my child and my delayed twin brother. This fact, unique to cloning, is bound to introduce confusion, but it is the sort of confusion that most people can overcome. The concern here is that because of the superficial confusion, a deeper confusing occurs, a confusing of the very meaning of human relationships as they are intended. At this point, of course, the argument ceases to be merely biological (a clone is an asynchronous twin, a confusing novelty) or psychological (being a teenager is hard enough already) and becomes theological (cloning confounds normative order). "Cloning oneself could only distort the distinctive, though integrally related, familial roles by negating the boundaries separating and delimiting the roles of parent, child, and sibling."[1] The argument thus depends upon a whole set of theological assumptions about God, creation, and the right ordering of human relationships in family.

Protestants, of course, disagree sharply with each other on the theological meaning of the family. Some Protestants, basing their argument in part on the Bible, oppose gay marriage and diversity in family forms. For them, cloning is just one more modern assault on the integrity of a God-defined institution. They believe that scripture holds up the enduring norm and model for human relatedness in the story of Adam and Eve, the first human pair who procreate sexually. Others argue that the story of Adam and Eve is non-normative, that for Christians the only truly normative human life is that of Jesus, who apparently did not have children or marry. Furthermore, the Bible is full of stories of people blessed of God in spite of the diverging forms of their family life, suggesting that there really is no single form for family, so there may be nothing theological for cloning to threaten. Whether God's blessing is to be found in life and in human relationships depends upon something other than their form or order.

Even if the theological assumption about a normative order for the human family is granted as valid, however, two additional points should be made. First, suppose cloning does introduce confusion into familial relationships, for instance by making me father and twin at once. Why must genetics be the overriding factor in defining or confusing relationships? Adoptive parents can be real parents for children with whom they have no genetic tie. They do not allow the genes to tell them not to love or parent the child. Is it not possible for cloning parents to recognize that even though they also deviate from the usual pattern of the parent–child genetic relationship, they too can choose to love as a parent while basing their parental authority on their years rather than on their genes? Granted, there may be moments when parent or clone might think that shared genes convey a mystically shared awareness of each other's thoughts or intentions. But more likely the cloned adolescent, like most adolescents, will believe that her mother, regardless of the percentage of shared DNA, does not understand her at all. The second additional point is that, even if one found the objection about confusion to be serious, it must be acknowledged that it simply does not apply at all if the genetic source of the clone is someone other than the parents.

Because of these limitations, the first objection does not justify a ban on cloning. At most, this objection serves as a point to be considered about the meaning of familial relationships and all that might confuse us in honoring them, and perhaps future would-be cloners should ask themselves whether they are likely to find this too confusing, or how they can take steps to avoid problems along these lines.

Cloning and Objectification

The second objection to cloning is that it has the potential to objectify or dehumanize the cloned child. If the existence and identity of the child are determined to a significant degree by technological choice and not by the uncontrolled outflowing of human love, then at least in relationship with parents the child is consigned to an inherently inferior status, as a product rather than a person. Such inferiority is not to be confused with the dependence of a child upon parents in early life, for the inferiority caused by technological control determines the relational identity of the child. The cloned child is made by technology at the service of human will, which is perfectly appropriate for any of life's projects or purchases but not for the engendering of equals. Humans must not attempt to make humans, for if they do, maker and artifact will not encounter each other as equally human.

In other words, cloning might introduce too great an asymmetry into the relationship between cloning parents and cloned child—an interesting irony because in one respect, the first objection is that there is too little difference between cloning parents their children. Now with the second objection, the opposite concern arises. Here the concern is that there is too great a difference between cloner and clone, a great gulf between designer and object. In other words, with cloning, what is most like me (my twin) is paradoxically least like me (my artifact).

Some who offer this objection are willing to apply it to reproductive technology across the board. Others see cloning as a special case, uniquely problematic among reproductive technologies, although it is difficult for them to point out exactly what distinguishes cloning in a religiously significant way. As a general rule, Protestants accept or at least are willing to tolerate reproductive technology. If reproductive technology is accepted but cloning opposed, there must be some religiously or morally significant difference between them. It is difficult to state clearly where the difference lies, which undermines the coherence of this objection.

One way to try to specify the difference between cloning and other reproductive technologies is to point to the fact that in nuclear transfer, the entire nuclear or chromosomal DNA of the embryo is determined or more precisely selected by technology. Other reproductive technologies retain an element of chance, valued in this case simply because it is a limit of control or specification. But the cloning process begins with the selection of a phenotype, an existing individual whose traits can be observed and then chosen, and whose nuclear DNA is then transferred to start a new life. It is this unique feature of cloning that raises the bar of technology too far, permitting too much control of one human being over another. After all, people do not just produce a child by cloning. They clone this or that genotype in the expectation of getting this or that type of child.

With cloning, "the entire genetic composition of a new individual is not random but has been chosen, controlled, and predetermined by another human being. It is this act of human control, I believe, that makes human reproductive cloning inherently wrong."[2] This control, once exercised, cannot be let go. Even though nothing else is different about the child, the fact of being cloned is a permanent difference. Whether or not anyone will ever actually experience this difference as a problem, as an assault on their dignity or as consignment to an inherently inferior status in life, cannot be known in advance. What can be known is that the fact upon which such feelings of inferiority might rest is a circumstance that cannot be reversed. There is, furthermore, no satisfying reason to think that being cloned or being otherwise created by technology will not be used to establish a future equivalent of racism or, perhaps, a form of self-depreciation.

What we must ponder is how the fact of being cloned might be interpreted by the cloners, the clones, and society as a whole. We should distinguish here between the effects of cloning on the parent–child relationship and its effects on the society–child relationship. In respect to the parent–child relationship, the child's creation by cloning may indeed introduce an enduring asymmetry, forever putting the child in an inferior status as object. In respect to the society–child relationship, the concern for asymmetry is lessened but not entirely missing. Cloning requires many collaborators, so it is not just the parents who clone. Medical and technical assistants are involved, as are larger institutions, even society as a whole and the nation, which fund, regulate, and legitimize these institutions and professions. So the asymmetry and possible inferiority that a cloned child might experience, while centered in the relationship with the parents, is socially inescapable. "They"—as in *they cloned me*—will have many referents.

But it must also be recognized that in comparing this situation to other forms of discrimination, which base theories of inferiority on some biological marker, cloning per se does not provide any such marker. Assuming that cloning can someday be done with complete safety, the only unique biological fact about the cloned child is the cloning itself, the unusual way the embryo is created. Will this fact really serve as a basis for discrimination or for feelings of inferiority? Will it be a problem for cloned children to know that they are clones? Anyone thinking so might advocate hiding this fact from them, but doing so might mean withholding medically valuable information, not to mention setting up a complex structure of deception that is likely to unravel.

A slightly different way to state this second objection is to say that cloning imposes an unfair burden of expectation upon the clone. Clones might wonder *why* they were cloned. Their parents will no doubt reassure them, saying something like what adopting parents say, that out of all the children, we chose you. But where cloning is concerned, that answer is slightly inaccurate and quite possibly troubling, for it is not the cloned child that is chosen but the clone's original, the previous human life with the same chromosomal DNA. Now of course, genes are not destiny, and we must take seriously all the criticism of genetic determinism. But that is quite beside the point, we can imagine, for the cloned child, who cannot face down the unsettling thought that *I was created to be like someone else*. Of course, we cannot predict whether a child will find this mildly interesting or deeply offensive, whether it's merely an innocuous side effect of cloning's novelty or a real and predictable harm, a kind of psychological injustice unfairly visited upon clones.

Even if one finds this objection generally convincing, it should be clear that it applies at most to some but not all cases in which cloning might be used. It applies best when the clone's original is known and selected for traits that the cloners wish to see repeated in the clone. It applies least when the clone's original is largely unknown. Consider for instance the parents of a very young child who suddenly dies. They might turn to cloning not to select the dead child's phenotype, which they hardly know, but merely to have a chance at a pregnancy of their own. Perhaps they face infertility problems and cloning is the only way to create a child of their own. In that case, they would rather not clone, but they have no alternative other than not having a genetically related child. In such cases it is hard to accuse these parents of engineering an object or burdening their child.

At the same time, at least under other circumstances, the objection does invite soul-searching by prospective parents who might consider cloning. Christian parents, especially, might reflect on whether they are seeking to create a child to love or to design a child worth loving.

Cloning and Grief

The final objection is also limited in its application, but for Christians it may be the most sobering of all our concerns. It is also the most difficult to communicate publicly because it rests largely on religious beliefs that are central to Christian faith and yet somewhat mystifying to Christians themselves, beliefs having to do with the hope of the resurrection of the dead. We can begin to think about this concern, however, by focusing on the common human experience of grief. The loss of a child is devastating beyond words, and who would fail to sympathize with grieving parents who might consider cloning their child? If we let grieving pet owners use cloning, how can we be less sympathetic to grieving parents?

Hearing this, Christians are likely to feel themselves sharply torn in their intuitions. As much as anyone, they will sympathize with those who grieve. But they will find themselves troubled by the suggestion that grief could be assuaged by technology, most of all by a technology that seems on the surface to offer a way to circumvent the loss itself, as if it could reverse grief by replacing what is lost.

Defenders of cloning will be quick to say that any suggestion that cloning replaces a dead child is simply bad science, and on that much they're right. But that merely adds weight to the objection, for precisely because cloning cannot replace the dead child, it should not be used in the context of grief when feelings are fragile and desires easily misled. The irrepressible yearning to reverse the death, to have the child back, to undo the tragic moment or find a cure for the incurable disease, all these desires welling up in the hearts of grieving parents cannot help but support the delusion that is at once comforting and disturbing: *cloning will bring my child back*. But what a temptation, to think that grief can be cut short, expedited, or removed altogether.

Our focus here, however, is not what Christians will tolerate for others but what they will accept and bless for themselves, either individually or within the context of the congregational community. We cannot predict how Protestants will respond when cloning is fully developed, but it is likely that some will consider cloning during grief. Perhaps upon reflection they will back away from cloning for fear that somehow it dishonors the dead. Of course, it does not replace the dead, but it treats them as replaceable. This might be acceptable for our pets but not for our children. From early life, children possess an irreplaceable and incommunicable uniqueness, their individual personhood, and anything that treats them as a replaceable cipher demeans this individuality. Just as we are aghast at the thought of replacing a living child by dismissing her from our family and bringing in another to sit in her place, so we would not seek to replace a dead child, for death does not diminish the irreplaceability. When the irreplaceable is lost, grief is the only response.

For Christians, experiences of death and grief are shaped by hope of resurrection. It is important to recognize that resurrection is not resuscitation, as if in the distant future the dead were simply brought back to life. If that were so, then cloning would compete here as a kind of surrogate resurrection. But resurrection is most of all a radical transformation of present existence into something wholly unknown, into a new creation of a spiritual nature. In fact, the contrast between cloning and resurrection is stark and twofold: Cloning creates a new individual in the same kind of body; resurrection restores the original individual in a different kind of body. Biology insists that cloning does not bring back the dead, and Christian theology maintains that it surely does not resurrect them. In fact, Christians would recognize that if they were to clone their dead child, the clone will die and in the end, both children—the original and the clone—will be raised. That thought alone might deter most Christians from cloning in grief.

Even so, in the turmoil of grief and in its ensuing confusions, cloning might be seen by some Christians as bringing back the dead and therefore as a kind of rival resurrection, a technological substitute. But this will surely bring a crisis of faith, as if trust is being transferred from God to technology. For Christians, God is the power of the future who in the end transforms all things and makes them new. What does it mean for a Christian to resort to cloning in grief, and to do so in the deluded but intense hope that cloning restores the dead to life? I think that Christians will reject this and see other Christians who do it as turning from their trust in God. To resort to cloning in the belief that cloning brings back the dead is to trust a power other than God to take care of the future. Cloning to restore the dead is a refusal to await the resurrection of the dead. It is an act of taking the future into our own hands rather than trusting in God. In that sense, cloning to restore the dead is an act of unfaith or unbelieving, an act of untrusting by refusing to wait for God to do what God alone can do.

When cloning becomes safe and available, will these concerns actually prevent Protestants from using it when, for instance, they lose a young child? Probably not. In fact, my own view is that these objections are all limited in their application and do not provide the basis for a categorical ban, even one for Christians to impose only on themselves. The mark of living as a Christian in today's world is not found in refusing to use some technologies but in learning the art of spiritual self-examination and in discovering how to use all things wisely.

Notes

[1] Brent Waters, "One Flesh: Cloning, Procreation, and the Family," in Ronald Cole-Turner (ed.), *Human Cloning: Religious Responses* (Louisville: Westminster John Knox Press, 1997), pp. 84–85.

[2] Donald M. Bruce, "Ethics Keeping Pace with Technology," in Ronald Cole-Turner (ed.), *Beyond Cloning: Religion and the Remaking of Humanity* (Harrisburg, PA: Trinity Press International, 2001), p. 40.

A Catholic Perspective on Stem Cell Research

Richard Doerflinger

Testimony of Richard Doerflinger, March 12, 2003, before the House Health and Government Operations Committee of the Maryland Legislature regarding stem-cell research and the donation of certain tissue for research purposes

I am Richard M. Doerflinger, Deputy Director of the Secretariat for Pro-Life Activities at the U.S. Conference of Catholic Bishops in Washington D.C. I also serve as Adjunct Fellow in Bioethics and Public Policy at the National Catholic Bioethics Center in Boston. I am a Maryland resident, having lived here (in Mount Rainier, then in Silver Spring) for twenty-two years.

I have been asked by the Maryland Catholic Conference to present testimony today on House Bill 482, on "Stem Cell Research." I am familiar with the bill because similar language is being marketed simultaneously in various states by certain elements of the biotechnology industry.

Regarding any bill it is a good idea to ask why it is needed. What does it approve or legalize that is not already being done legally *without* the bill's enactment? After all, in American law everything is allowed unless it is expressly prohibited. So what is currently illegal, or at least legally or morally controversial, that needs to receive this legislative endorsement?

What HB 482 is not needed for

This bill is certainly not needed in order to allow adult stem cell research, which is completely legal and eligible for federal funding. Nor is it needed in order to allow the use of fetal tissue (including "embryonic germ cells" harvested from unborn children who are aborted at around 8 weeks of gestation). Such research is not only legal, but is currently conducted at Johns Hopkins University and (with certain conditions) is also eligible for federal funding. Nor is it even needed in order to allow use of embryonic stem cells obtained by destroying one-week-old human embryos who are considered to be "in excess of clinical need" donated at fertility clinics. Such research is legal in Maryland and most other states, and even some aspects of this research (i.e., research on stem cell lines that were derived from embryos before August 9, 2001) are eligible for federal funding. And the bill is not needed to set legal limits on the selling of "embryonic or cadaveric fetal tissue," since exactly these same limits have already been part of *federal* law for a decade (and incidentally, have shown themselves to be inadequate and ineffectual). See 42 USC ?289g-2.

What HB 482 approves

The central purpose of this bill, then, must be to approve that part of "stem cell research" that is either illegal, or so widely condemned that researchers might fear to pursue it without the legislature's seal of approval. If the legislature approves this bill, it will do so to ensure that the following is conducted in the state of Maryland:

1. **Research in which human embryos are *specially created* by in vitro fertilization solely in order to be**

destroyed for research purposes. A program of this kind was begun at a fertility clinic in Norfolk, Virginia in 2001, but was discontinued after it provoked nearly universal moral outrage from lawmakers and others. Even to many people who accept the idea of research on already existing "spare" embryos who may otherwise be discarded, the idea of creating early human lives solely to destroy them takes a giant step toward treating life as a mere instrument. A Maryland researcher may well want to have the advance approval of the state legislature before beginning such a project here.

2. **Fetal tissue research that violates the ethical standards required for federally funded research.** If it did not go beyond the federal funding standards it would have no need of official encouragement from the state. This could include projects in which:

- Researchers desiring fetal tissue perform the abortions themselves, or introduce changes in the timing or method of abortion to produce more usable tissue;
- Women are persuaded to have abortions precisely in order to provide transplant tissue for themselves or their loved ones.
- Women are encouraged to become pregnant so that such abortions can be performed in order to obtain fetal tissue for themselves or their loved ones.

3. **Research using "somatic cell nuclear transplantation" (cloning) to produce human embryos *and* fetuses *and* born children as sources of cells and tissues.** This sounds like a nightmarish science fiction scenario, but it is exactly what the bill says:

> "The General Assembly declares that it is the policy of the state that research involving the derivation and use of human embryonic stem cells, human embryonic germ cells, and human adult stem cells from any source, including somatic cell nuclear transplantation, shall be allowed."

The idea of using cloning to mass-produce and harvest human embryos as nothing more than agricultural products is horrific enough (though it is not currently illegal in the state of Maryland). The idea of extending this into the "farming" of fetal and newborn humans for spare parts, however, is the research area that most needs legislative endorsement if scientists are to have the nerve to pursue it.

Technically such research does not seem to be illegal now—that is, Maryland has no law against any kind of human cloning, and effectively no limits on abortion. And as long as parental consent is obtained, and the cell harvesting is not harmful to the child, there is no law against cloning and delivering a child solely to obtain his or her bone-marrow stem cells to treat the adult whose genetic "copy" the child was created to be. But understandably, researchers will not pursue these avenues without official approval, because almost everyone on earth would see them as morally monstrous.

Why bills like HB 482 are being recommended to state legislatures

The language of this bill is no accident, at least from the viewpoint of outside groups supporting it.. A similar bill (SB 1909, AB 2840) was considered in New Jersey this year—it banned "human cloning," but then said that one is not guilty of "cloning" unless one develops the cloned human *through* the embryonic, fetal *and newborn* stages to produce a human "individual." The bill actually was approved by one chamber of the legislature, then ultimately withdrawn when legislators finally realized what it does. In New York a bill (A. 6249) has been introduced to allow human cloning, and the gestating of cloned embryos through the fetal stage, *as long as this is not done for the purpose of producing a live birth.*

Why is this being proposed now? It is done because certain interests in the biotechnology industry, fixated on the pursuit of human cloning for biomedical research, are afraid that their original model for "therapeutic cloning" may not work. The original idea was to use cloning to produce one-week-old embryos, who would then be destroyed for their "embryonic stem cells" which will be a perfect genetic match to the patient who donated genetic material for the cloning procedure. But now problems have emerged.

It turns out that "therapeutic cloning" is not working well, even in animals. Embryonic stem cells are too difficult to maintain, too uncontrollable, too likely to turn into lethal tumors in animals' bodies. There are only two animal studies suggesting therapeutic benefits from cells that originated from cloning. One, designed to provide new kidney tissue, required gestating the cloned cow embryo in a uterus and then aborting it to obtain *fetal* kidney tissue (Lanza et al., 20 *Nature Biotechnology* 689–96, July 2002). The other, designed to correct a genetically-based immune deficiency, required taking the new mouse (produced by cloning and genetic modification) to the *newborn* stage and harvesting its *adult* stem cells to treat the original mouse (Rideout et al., 109 *Cell* 17–27, April 5, 2002). These new state bills on cloning are designed to keep researchers' options open, to make it possible to transfer these animal models to human use.

This means, of course, that in approving this bill Maryland would be marking out a path that is rejected by the vast majority of Americans, and apparently by every single member of Congress. For there is no member of Congress who supports allowing cloned embryos to be placed in a womb and gestated to the fetal or newborn stage (called by some "reproductive cloning"). Some members of Congress, including some of the Maryland congressional delegation, say they want to allow use of cloning to make embryos who will be destroyed for their stem cells at an early stage in the laboratory—but the introduction of bills like these illustrates that the research community is already abandoning that stance and is moving on to the next step.

Ethical limits?

Doesn't the bill prevent these horrors by its call for attention to ethical limits? No, it does not. There is vague talk of "consideration of the ethical concerns regarding this research," but no actual limits. Surely no court would find that the legislature, having officially *approved* cloning to produce embryos, fetuses and newborns, intends to have this clear mandate overridden by someone's "ethical concerns" after the fact. And the bill speaks of approval by an Institutional Review Board, but provides no standards by which the IRB could judge an experiment. Usually, in approving even privately funded research, IRBs follow the standards for federally funded research by default. That cannot be done here, because research that involves doing abortions to obtain fetal tissue, destroying embryos for their stem cells, and performing human cloning is completely ineligible for federal funding. So there are no standards. The IRBs will have to fly us into the Brave New World of cloning and body farming by the seat of their pants.

In so doing, the IRBs would be able to glean the following ethical guidance from the language of the bill itself:

> "Stem cell research could lead to unprecedented treatments and potential cures" for a wide array of diseases.

Incidentally, the truth of this statement is highly questionable in the case of embryonic stem cell research and "therapeutic cloning," neither of which is anywhere near providing any treatment for humans; the statement about producing cures would be true of many aspects of adult stem cell research, but that is already legal and federally funded and in no need of help from this bill.

> "The United States and the State of Maryland have historically fostered open scientific inquiry."

This is a half-truth. Our federal and state governments have insisted that scientific inquiry not be pursued if it will mistreat human subjects. "Historically," there have been lapses in which the lust for medical progress was allowed to outweigh ethical concerns, but these are the darkest pages of our national history. Maryland, in fact, has been in the forefront of states that insist on clear ethical and legal limits to medical research, and its courts have found that researchers can be sued for negligence when they give insufficient attention to the safety and well-being of children who are too young to consent to research. See *Grimes v. Kennedy Krieger Institute*, 782 A.2d 807 (Md. 2001). Incidentally, Maryland's highest court found in this decision that the IRB at Johns Hopkins University had "abdicated" its responsibility to protect children from research risks, and had shown that it "was willing to aid researchers in getting around federal regulations designed to protect children used as subjects in nontherapeutic research." No one who has read this decision will want to entrust all ethically controversial research decisions solely to IRBs.

> Maryland's biotechnology industry is a gold mine for the state, and this situation would be endangered by "limitations" on stem cell research. Therefore, "public policy on stem cell research must *balance* ethical and medical considerations."

This statement of principle sounds nice and moderate, until one gives it a moment's thought. Since the Nuremberg Code, society has insisted that the lure of medical progress must *never* be misused to outweigh ethics: "Concern for the interests of the subject must always prevail over the interest of science and society" (World Medical Association,

Declaration of Helsinki (1975), I.5). Or as a survivor of Dr. Josef Mengele's notorious "twin" experiments has said: "Human dignity and human life are more important than any advance in science or medicine" (Eva Mozes-Kor, quoted in A. Caplan, *When Medicine Went Mad* (1992), p. 7).

If researchers are to "balance" ethics and medicine, that means one can pursue *even research that one knows is unethical*—if the potential medical benefit is great enough. The researchers who conducted the Tuskegee syphilis experiment on African-American men, who deliberately infected retarded children with the hepatitis virus at the Willowbrook home, who approved the Cold War radiation experiments on unsuspecting American civilians, who allowed children to be exposed to lead poisoning in the research condemned two years ago by the Maryland Court of Appeals—these researchers "balanced" ethics and progress, and "progress" won. I urge the state of Maryland not to pursue such "balance."

Finally, a word about the bill's attempt to ban the selling of fetal organs and tissues. I'm sure this part of the bill is well meant, but we know in advance it will be ineffectual. It tracks the language of a federal ban that has been in place since 1993: Such tissues and organs may not be given for "valuable consideration," but a service fee for obtaining the tissue, shipping and handling etc., is allowed. Under this language, fetal tissue distributors have been able to advertise the availability of various fetal organs and tissues for pay, publish "price lists," and in brief do everything involved in selling body parts—as long as they *call* it a "service fee." If Maryland wants to prepare legislation to prevent such abuse, that may be a laudable idea—but we know this language is not the way to do it.

Conclusion

My testimony will sound largely negative—and it is negative regarding this bill, which is a moral and medical horror. But I want to end with a message of hope. In recent years, enormously promising new treatments have begun to emerge from medical research. In clinical trials, patients have been successfully treated for Parkinson's disease, sickle-cell anemia, thalassemia, Type I diabetes, severe combined immune deficiency, corneal damage, heart disease, bone and cartilage injury, and so on. Early trials in treating patients with chronic spinal cord injury have shown progress in regaining sensation and movement. All these trials used *adult* stem cells, other adult cells obtained ethically, or sources such as umbilical cord blood. Some of this progress has even come from researchers in Maryland. No treatment now in trials, or on the horizon, comes from embryonic stem cells or from so-called "therapeutic cloning." By focusing on these most controversial and most speculative approaches, our resources and attention may well be diverted away from the promising treatments now on the brink of helping millions of patients—and we will actually slow down the medical progress that patients deserve.

We need not set aside human dignity, or "balance" ethics against the desire for progress—a balancing act in which ethics so often seems to lose. We can fully respect and promote sound ethics and promising medicine together, for there is no conflict between them. I urge the committee to begin this journey to ethically responsible medical progress by defeating HB 482.

In God's Garden: Creation and Cloning in Jewish Thought

Jonathan R. Cohen

The possibility of cloning human beings challenges Western beliefs about creation and our relationship to God. If we understand God as the Creator and creation as a completed act, cloning will be a transgression. If, however, we understand God as the Power of Creation and creation as a transformative process, we may find a role for human participation, sharing that power as beings created in the image of God

Some scientific revolutions change what people believe about the world. The Copernican and Darwinian revolutions, while not significantly changing what people could then use science to achieve, forced people to re-examine their understanding of the universe, of humanity and its place in the universe, and of God's agency within the universe. Other scientific revolutions, like Faraday and Maxwell's work on the physics of electric fields, pose little challenge to fundamental beliefs but dramatically change society through their technological application. We are now in the midst of a genetic revolution that may both profoundly influence our beliefs and dramatically change how society functions. Will designing our offspring someday be as easy and common as "cut and paste" on a word processor? Are we on the cusp of an evolutionary advance toward being an "autocreative" species, or in attempting to "play God" has our hubris reached its zenith? Such are the questions we face.

My purpose here is to explore how the genetic revolution could affect our beliefs. My strategy is to examine several challenges that human cloning and, to a lesser extent, genetic engineering raise for certain basic Jewish beliefs—though by no means exclusively Jewish beliefs—about humans and God, using as a lens for my thoughts the Biblical account of creation presented in the first few chapters of Genesis.[1] Not only are many basic Jewish beliefs about humans and God embedded in that account, but even if one believes that Genesis is inaccurate as a literal account of the world's creation, or even if one does not believe in God, Genesis provides an excellent framework for addressing some of the existential challenges posed by the genetic revolution. Although I address challenges human cloning presents for certain basic Jewish beliefs, such Biblically-rooted beliefs influence much Christian and Western thought. Moreover, such beliefs play an important part in developing public policy toward human cloning. For example, the National Bioethics Advisory Commission devoted roughly one quarter of its report on human cloning to religious views, focusing in particular on the Biblical account of creation.

The Biblical account is open to two different interpretations, of creation as a completed act and creation as a transformative process, which carry quite different implications for human cloning. Understanding creation in these different ways suggests different answers to how human cloning might impinge on our beliefs about the worth of a human life, about God's role as Creator and Sovereign, and about how meaning can be found in a human life. I want to suggest that if its implications are properly understood, human cloning can be integrated with many of our basic beliefs and can encourage us to view God's act of creation as a transformative process.

This is not to advocate that human cloning be permitted—that is a very different question. However, the possibility of human cloning challenges our beliefs irrespective of whether we ultimately permit or ban such practice, or of whether human cloning actually occurs.

A note before I begin. Although for simplicity I use terms such as "Jewish beliefs" and "Jewish thought," I do not mean to suggest that all Jews do hold or should hold similar beliefs about these topics or that Judaism requires one to hold a particular view about these topics. Indeed, I pretend no special expertise in Jewish thought, but speak as a lay Jew who seeks to make some existential "sense" out of the possibility of human cloning.

Completed Act or Transformative Process?

The Bible begins, *"Bereshit bara Elohim et hashamayim v'et haaretz..."* It is a mysterious phrase. Under one common interpretation, it is translated as a declarative sentence—"In the beginning, God created the heavens and the earth." The import is that God created the universe out of nothing, and essentially all at once. In this reading, God engages in two primary activities in Genesis: bringing into existence various elements, and dividing them from one another, each to have a distinct role. Light is separated from darkness; land is separated from sea; birds are to fly in the sky while fish are to swim in the sea. The patterns of reproduction also appear Divinely set. Each form of vegetation is to produce offspring of its own type, and there are two genders of humans, with reproduction to occur through the union of a male and a female.[2] We are also told that by the seventh day, "The heavens and the earth and all their hosts were complete" (Gen. 2:1). The structure of the universe had been set. Creation is essentially completed.

From such an interpretation, an argument arises: if God created the structure of our world, who are we to tamper with it? Further, the Bible describes humans as created "in God's image" (Gen. 1: 27). If "in God's image" means "after God's likeness," how could that likeness be improved upon? If "in God's image" means "in accordance with God's plan," who are we to create a better plan?

If God's stamp in creating the world and the life within it alone does not lead one to think that the structure of the world should be left as is, other sources might. The Bible depicts transgressing the boundaries God gave as the paradigm of sin. Eating the forbidden fruit leads to the expulsion from Eden, and the commingling of divine and human beings precedes the flood. Sex between humans and animals is prohibited, as is sex between two men (Lev. 18: 22–24; 20: 12, 15–16). So too crossbreeding animals and planting a field with different types of seed (Lev. 19: 19; Deut. 22: 9). The concern for nature's structure is even applied to clothing: it is forbidden to construct a garment of both linen and wool (Lev. 19: 19, Deut. 22: 12). Conversely, many Biblical passages, especially in Leviticus, indicate that holiness can be found by respecting boundaries. As Mary Douglas has argued, "Holiness requires that individuals shall conform to the class to which they belong. And holiness requires that different classes of things shall not be confused."[3]

Against such a reading of the Bible, human cloning would be wrong. Although human cloning would not bring into existence a new species-a potential criticism of transgenic activities such as making hybrid plants or animals—it would transgress the structure of sexual reproduction that God created. Human cloning supplants the structure by which God designed humans to reproduce. Our current genetic quandary might be cast as a second fall from Eden. Driven by our lust for Godlike power, we have picked of the tempting fruit of the tree of genetic knowledge. We ought not to use that knowledge to pervert the structure of the world.[4]

Like most great literature, however, the Biblical account of creation can be interpreted in different ways. A second, contrasting interpretation of Genesis may be offered that has quite different implications for how we view human cloning. Often the opening phrase of Genesis is translated not as a declarative sentence, but as a constructive clause—"When God began creating the heavens and the earth..." or "At the beginning of God's creation of the heavens and the earth..."[5] So construed, creation may be seen not as a completed act cast in a particular structure, but as a transformative process.[6] In this interpretation, the miracle of creation is not the specific world God produced, but rather God's moving the world from a chaotic nothingness to an ordered, light-filled, life-bearing place. Further, one might point to the Bible's repeated emphasis that God created things called "good" and "very good." Put differently, the miracle was that God improved what existed. Good purpose, rather than a particular form, lies at the heart of creation.[7]

If God is seen as Creator, and if we are created in the image of God, then might we not have a role to play as creators ourselves?[8] Abraham is viewed as praiseworthy when, exercising an independent conception of what is moral, he argues with God over the fate of Sodom and Gomorrah.[9] Might we be praiseworthy if we put our technology to work to pursue our independently formed conception of the good?[10] People get sick naturally, and yet most feel that medical intervention to aid the sick is morally permissible, perhaps even obligatory, as in Jewish law. God, rather than nature, is to be worshipped.

If creation is a transformative process of bettering our world in which humans are to play a part, then in assessing human cloning the normative focus would turn to whether we use human cloning to do good or evil. As Rabbi Elliot Dorff has written, "Cloning, like all other technologies, is morally neutral. Its moral valence depends on how we use it."[11] It is in this respect like our use of drugs: when used to improve health, they are a blessing; when taken by addicts, a curse. And some think that the use of human cloning would sometimes be merited, even obliged. Rabbi Moshe Tendler has declared, "Show me a young man who is sterile, whose family was wiped out in the Holocaust, and [who] is the last of a genetic line [and] I would certainly done him."[12] Other candidates include more common cases of infertility, such as parents who could not otherwise reproduce and want to done their recently deceased newborn, or of saving someone's life, such as cloning an infant who has suffered severe kidney damage in the hope that the clone might someday willingly donate a kidney to the clonee.[13]

In sum, different interpretations of Genesis have quite different implications for how we judge human cloning. If we believe structuring our world a particular way lies at the heart of God's creation, then we will likely view human cloning as transgressing that structure. In contrast, if we believe that transforming what exists for the better lies at the heart of creation, then our view of human cloning will likely depend on whether we use human cloning to accomplish good or evil.

Cloning and Humanity

Central to the Western conception of human nature is the Biblical view that we were created by God and "in God's image." The idea has been historically as well as philosophically important, for it has done much to protect and elevate the status of humans. Yet the genetic revolution, and especially the possibility of human cloning, deeply challenges that view. While humans have long produced other humans through traditional reproduction, they have never been able to control the exact genetic structure of their offspring. Traditional conception has always involved much randomness and uncertainty. Seeing God's hand in the uncertain and mysterious is relatively easy; seeing God's hand in what we can control may be difficult. Cloning and genetic engineering offer the prospect of removing that randomness and uncertainty, and so threaten to undermine the belief that humans are created by God, in God's image.

Commenting on the Biblical account of creation, the Mishnaic Rabbis (c. 200 CE) explained:

> For this reason was man created alone, to teach you that whoever destroys a single soul of Israel, Scripture imputes (guilt) to him as though he had destroyed a complete world; and whosoever preserves a single soul of Israel, Scripture ascribes (merit) to him as though he had preserved a complete world. Furthermore, (he was created alone) for the sake of peace among men, that one might not say to his fellow, "My father was greater than thine", ...[and] to proclaim the greatness of The Holy One, Blessed be He: for if a man strikes many coins from one mould, they all resemble one another, but The Supreme King of Kings, The Holy One, Blessed be He, fashioned every man in the stamp of the first man, and yet not one of them resembles his fellow.[14]

The commentary exemplifies the traditional Jewish view that three central values are imputed by the Biblical account of creation to every human life: pricelessness, uniqueness, and equality. Arguably, cloning could undermine each of these values.

Consider pricelessness tint. Belief in the pricelessness of human life seems to fall naturally out of the belief that humans are created in God's image, for if each of us is created in God's image, then each of us is of infinite worth, indeed is sacred. The worry is that if we clone our offspring, then rather than seeing them as created in the Divine image, we might come to see them as mere objects of production, genetically replaceable like other products. The art market provides analogies: An original oil painting is typically much more valuable than copies of it, and objects that can be readily duplicated, such as photographs, usually sell for far less than those that cannot. In economic language, cloning increases the potential "supply" of each of us and so might cause our value to decline.

Yet this fear should not be overstated. If someone were cloned a thousand times over, perhaps it would be hard to see each as a priceless being. But such use seems unlikely. In contrast, if cloning were used to make only one or two "copies" of a person, maintaining the belief that each has infinite worth would be much easier. Few would argue that natural genetic twins have diminished worth.[15]

As interpreted by the Mishnaic Rabbis, the fact that God began by creating not a group of people but the individual Adam also shows that every human being is unique.[16] To this day, a belief in individual uniqueness has played an important part in Jewish thought. As was expressed by Rabbi Zusya of Anipol shortly before his death, "In the world to come I shall not be asked: 'Why were you not Moses?' Rather I shall be asked: 'Why were you not Zusya?' "[17] Martin Buber also put great weight on the concept of human uniqueness:

> Every person born into this world represents something new, something that never existed before, something original and unique... Every man's foremost task is the actualization of his unique, unprecedented and never-recurring potentialities, and not the repetition of something that another, and be it even the greatest, has already achieved. (p. 17)

To many (myself included), the thought of being cloned is frightening. Indeed, the very thought that one could be cloned may be disturbing. If I can be copied, what is so special about me?

Of course, as many have pointed out, two clones would not really be identical. Raised in different environments, perhaps at different times, they will become different people. Even if physically identical, each will have a different character—a different soul. Yet for many, this observation only ducks the question; even if a clone will not be in all ways identical to the one cloned, he or she will be similar in many ways and identical in one fundamental way—namely, in having the same genetic composition.

Ultimately, cloning challenges us to consider how important our genetic structure is to our sense of self. Specifically, it challenges us to consider to what extent a person is more than a physical being-or, to the degree that behavior is genetically influenced, more than just a set of particular behaviors. The less one's sense of identity is based on physical being, the less threatening cloning becomes. If when one looks in the mirror one sees only one's physical being, then a genetic duplicate might destroy one's sense of uniqueness.

Cloning also forces one to ask how important uniqueness is to one's sense of self. Why should one be a lesser person simply because there are copies of one? Contra Buber's view, perhaps a person's foremost task is not the actualization of his or her "unique, unprecedented and never-recurring potentialities," but simply the actualization of his or her potentialities, whether or not others possess those potentialities as well.

When first exposed to photography, some people refused to be photographed for fear that a photograph—an inanimate, two-dimensional copy—would "capture their souls." Over time, most of us have learned to tolerate the camera. For those who do not mind being photographed but are repulsed at the thought of being cloned, a useful thought experiment is to ask at what point our repulsion toward cloning begins. Would we be repulsed by an inanimate, three-dimensional copy—a statue? A three-dimensional copy that is mechanically animated? That is biologically animated? That can think?

The third lesson often tied in the Jewish tradition to the Biblical account of human creation is that of equality. If God initially created one person (Adam), and we are all descended from Adam, then we must all be equal.[18] Our common descent implies our equality.

What are the implications of cloning—and of genetic engineering—for equality? Would lesser people be squeezed out? Would we produce a basketball team of Michael Jordans or a university of Albert Einsteins? Would neo-Nazis produce their "master" race? Would we breed docile workers? Would the rich become genetically advantaged? Would we see people produced by genetic engineering as better or worse than those produced by traditional means?

Yet the challenge to equality, like the challenges to pricelessness and uniqueness, is also conceptually no greater than challenges we already face. Already there are significant genetic differences between people, and yet we view all people as equal. Already identical twins exist, and yet we view each as priceless and unique. Human choice, rather than genetic structure, has long determined the values we attach to human life.

Theological Implications

Just as the possibility of human cloning challenges our belief that humans are created in God's image, it also challenges our image of God as our Creator—our "Parent" or "Father." Of course, advances in genetic knowledge will not solve the great mystery of where the universe in toto came from, but because it suggests that we can "autocreate," it does seem likely to affect our own relationship to the creative power of God.

Again, it seems, we could turn to the transformative view

of creation, augmented by the view that we participate in the creation. And perhaps it can be admitted that we participate in the transformation. Perhaps, instead of seeing God as an agent who acted in the distant past, we will see God as the Power of Creation, and hold that we too share in that Power.[19] If asked whether we are "playing God" by engaging in human cloning, we might respond, "Yes, for God is in us too." We might even stress that creation lies not merely in changing the world, but in changing it for the better.

A similar point holds for our image of God as Sovereign—as "Ruler" or "King." Under the view that creation is a completed act, our autocreativity seems to usurp God's sovereignty. But if our understanding of God's sovereignty can parallel the changed understanding of our relationship to God's creative power, it need not. If creation involves changing the world for the better, not merely tampering with it, then we might see God's sovereignty as requiring the responsible exercise of the Godliness within ourselves.

Perhaps all this seems arrogant, but I do not find it so. Recognizing that responsibilities attach to the powers we have, and accepting those responsibilities, may form the basis of a more mature understanding of ourselves and of God and God's sovereignty. Children must eventually become adults, and works of art must stand on their own.

Death and the Search for Meaning

The Biblical account of creation concludes with expulsion. As punishment for eating the forbidden fruit of the tree of knowledge, Adam and Eve are expelled from Eden and blocked from eating the fruit of the tree of (eternal) life, lest they, like God, become immortal (Gen. 3: 22). Despite a presumed desire for immortality, death awaits, and Adam and Eve must live in its shadow.

Some might think that cloning (and perhaps other new

genetic technologies, such as those derived by replicating embryonic stem cells) offers a way to escape death—a way to make themselves "immortal." Through cloning a person might try to ensure that he or she does not really die; indeed, one might suppose it possible to achieve immortality by spawning a series of clones over time. Others might seek immortality of a different sort by producing multiple replicates all at once, so as to spread their genes as widely as possible and ensure their perpetuation in the human stock. Or one might try to preserve oneself by creating clones to supply genetically identical, but more youthful, body parts.

Of course, there are a variety of specific reasons that such attempts either must fail or are morally abhorrent. A general response is also available, however. As the author of Ecclesiastes asked, what point is there in living, if death awaits us all (Eccles. 3: 18–19)? Yet even if, *arguendo*, immortality or near immortality could be achieved through human cloning, rather than asking, "What point is there in living if death awaits?" a simpler question would remain, namely, "What point is there in living?"

The Biblical author was well acquainted with the human search for meaning and well aware that personal immortality can be a seductive, yet fruitless, goal in that search. In contradistinction to other Near Eastern religions concerned with death and immortality (as seen, for example, in Egyptian embalmment and mummification), the Bible strictly limited contact with the dead (Numbers 19: 11–16). The approach advocated in the Hebrew Bible is to live in connection with the Eternal in the life one leads, rather than to seek eternal life. Indeed, death was later taken as an impetus for spiritual growth. As the Psalmist expressed, "Teach us to number our days so that we may maintain a heart of wisdom" (Psalms 90: 12).

For many, the greatest human existential dilemma is not death but isolation and loneliness, and the greatest source of meaning comes from finding a mate and having children. In the Biblical narrative, shortly after Adam's creation but before Eve's creation—and before Adam's mortality or immortality is clearly established—God offers a rare comment on the human condition: "It is not good for man to be alone" (Gen. 2: 18). God then creates Eve to be Adam's counterpart.

Cloning would be a poor substitute for what is achieved through mating, understood in the richest sense, as involving bonding with another and sharing of life's joys and sorrows, and joining with that other to create an original and genetically intertwined life. By contrast, in cloning oneself one would be focused primarily on one's own life, trapped within one's own ego.

Perhaps the Biblical verb used to indicate sexual relations—*yodeah* (to know)—hints at the deep role that bonding with another person, including joining physically and genetically, may play in finding meaning in life. At its best, joining with another to create a child is an act not merely of reproduction, but of love. While there are many sources of meaning in life, it is hard to imagine one greater than love.

Wild Strains and Cultivars

A Jerusalem rabbi once shared with me a story that I have found helpful, indeed comforting, in thinking about the genetic revolution. I had asked this rabbi about raising my children, may I someday be so blessed, in the United States, versus raising them in Israel. Would one option provide a better life for them as Jews than the other? He responded with a story that he attributed to the Ba'al Shem Toy, the mythical, eighteenth-century founder of Hassidic Judaism. The Ba'al Shem Toy taught that there are two types of fruit in the world: fruit that grows in vineyards, and fruit that grows in the wild. Usually, fruit that grows in vineyards is large, shapely, tasty, and consistent. Fruit that grows in the wild often has blemishes or defects, and much of it is lost to insects and disease. However, it may be quite strong in flavor. How do these two types of fruit compare? Both are pleasing in God's eyes.

In time, we may well see a world in which many people will be cloned or genetically engineered, while others will be created through traditional means. Perhaps both will be pleasing in God's eyes.

Notes

1 For an overview of religious issues raised by the possibility of human cloning, see National Bioethics Advisory Commission, *Cloning Human Beings: Report and Recommendations of the National Bioethics Advisory Commission* (Rockville MD, 1997); Ethics and Theology: A Continuation of the National Discussion on Human Cloning: Hearing Before the Subcommittee on Public Health and Safety of the Senate Committee on Labor and Human Resources, 105th Cong. (1997). For analyses of genetic engineering from a Jewish perspective before the "Dolly" breakthrough, see Azriel Rosenfeld, "Judaism and Gene Design," and Fred Rosner, "Genetic Engineering and Judaism," both in *Jewish Bioethics*, ed. Fred Rosner and J. David Bleich (New York 1979), pp. 401–408 and 409–420, respectively. For Christian perspectives on human cloning, see Ronald Cole-Turner, ed., *Human Cloning: Religious Responses* (Louisville 1997).

2 Arguably, different accounts of how the genders arise can be found in the first and the second chapters of Genesis, which many have observed appear to provide not one but two accounts of creation. Traditional commentators have sought to reconcile these accounts (and thereby defend the view that the entire Bible is the word of God as transcribed by Moses). For one such recent work, see Joseph B. Soloveitchik, *The Lonely Man of Faith* (New York 1992). For the view that the Bible is composed of many documents, see E. A. Spieser, *Anchor Bible: Genesis* (New York 1962), pp. xx–xxii, 3–28.

3 Mary Douglas, "The Abominations of Leviticus," in *Purity and Danger: An Analysis of the Concepts of Pollution and Taboo* (London 1984), pp. 41–57, at 53.

4 A parallel argument can be made in evolutionary terms. The

genetic structure of our world evolved over billions of years into an interwoven and equilibrated system. While we often use science to modify nature, such modifications function within an existing evolutionary structure. In contrast, human cloning (and genetic engineering more generally) changes the very rules of the game of genetic evolution. Such a profound shift may shatter the entire system.

[5] See, for example, *Torah* (New York: Jewish Publication Society, 1962); Everett Fox, *Five Books of Moses* (New York 1995); and Robert Alter, *Genesis* (New York 1996). Under this second translation, a formless and void earth might, though need not necessarily, be supposed to have existed before God first acted by creating light.

[6] For a similar argument from a Christian perspective, see Ted Peters, *Playing God? Genetic Determinism and Human Freedom* (New York 1997), p. 14, distinguishing between *creatio ex nihilo* and *creatio continua*; and Philip Hefner, "The Evolution of the Created Co-Creator," in *Cosmos as Creation: Science and Theology in Consonance*, ed. Ted Peters (Nashville 1989), pp. 212–33.

[7] To explain the textual claim that by the seventh day God had completed the structure of the universe, various points might be made. For example, the text stresses that God rested from the work he had completed (Gen. 2: 2–3), but does not explicitly say that God did not engage in further work later. Indeed, the second chapter of Genesis proceeds to offer a second account of creation.

[8] When describing Divine creation, the Bible uses two verbs: bara (to bring into existence) and yatzar (to form or shape). However, when describing creation by humans, the Bible uses only one verb: yatzar. Some might feel that even if we humans should view ourselves as creators, we should limit ourselves to yatzar, a category that might exclude human cloning. The questions would then become: (1) What does one mean by bara and yatzar? and (2) Where on that spectrum do the various uses of genetic technology fall? I do not seek to resolve these questions here, but I do believe that how one resolves them depends in part on the extent to which one views creation as a completed act versus a transformative process.

[9] Arguing with God is a recurrent theme in Jewish thought. See Anson Laytner, *Arguing with God: A Jewish Tradition* (Northvale NJ, 1990).

[10] Some may ask whether creation by humans should be judged

differently from creation by God. If God created something, then we can presume that something to be good; but if humans create, we can make no such presumption, the argument would run. Yet the Biblical text supports a different view. The Bible suggests that the merits of God's creations must also be judged and cannot simply be deduced from their source or fully foreseen in advance. See Gen. 1: 3, 10, 12, 18, 21, 25, and 30, where God assesses God's own creations as "good" and "very good." See also Gen. 5: 5–13, especially 12, where God's destroys most of the world by flood after assessing the earth as corrupt. Similarly, one might think that the merits of human creation cannot be fully foreseen but must await later assessment.

[11] Elliot Dorff, "Human Cloning: A Jewish Perspective," Testimony before the National Bioethics Advisory Commission (14 March 1997), p. 5. Contrast J. M. Haas, Letter from the Pope John Center, submitted to the Nation Bioethics Advisory Commission (31 March 1997), p. 4.

[12] Rabbi Moshe Tendler, Testimony before the National Bioethics Advisory Commission (14 March 1997), 10–11, at 10.

[13] See James F. Childress, "The Challenges of Public Ethics: Reflections on NBAC's Report," *Hastings Center Report* 27, no. 5 (1997): 9–11, at 10.

[14] Babylonian Talmud: Seder Nezikin, vol. 1, tr. Isidore Epstein (London, 1935), pp. 233–234 (Sanhedrin 37a). Epstein points out that the term "of Israel" is omitted in some versions (p. 234, note 2). No doubt the editors of those versions were aware of the tension the phrase "of Israel" created with the ensuing verse proclaiming human equality.

[15] However, natural genetic twins, unlike clones, are not produced with the intention of achieving genetic identicalness.

[16] In one interpretation, God did not begin human creation with a single individual. See particularly Gen. 1: 26–28.

[17] Quoted in Martin Buber, *The Way of Man* (Chicago 1951), p. 18.

[18] Scholars debate whether the Biblical accounts of creation reflect men and women as equal. See Gen. 2: 16.

[19] For such an approach, see Mordecai M. Kaplan, *The Meaning of God in Modern Jewish Religion* (New York 1937), pp. 25–29, 51–57, and 62.

Human Clones and God's Trust: The Islamic View

Munawar Ahmad Anees

A specter haunts our future. It is the phantom of our phylogenetic past that has been decoded by science and used to replicate human life. What for all previous human cultures had been the unknown has come to be known through the act of human cloning. What are the implications of this audacious act by man? How should it be regarded by Muslims?

We are not looking here at simply another newfangled technique. Through the prism of replicating cells, we are peering into the past, and the future. Only three decades ago, the so-called genetic code was discovered as a sequence of four bases across the helical structure of DNA. Today, we are engaged in a gigantic endeavor to map the entire genetic makeup of man through the Human Genome Project. It is for biology what the Manhattan Project, which devised the nuclear bomb, was for physics.

Terrible secrets are being unlocked. Already in this short span of time many a taboo stands obliterated: artificial insemination by donor, *in vitro* fertilization and surrogate motherhood to name just a few variations on the theme.

Lest we baptize the first human clone with the fluid truths of postmodern casuistry, that is to say clever but false arguments, let us be honest about what this foreshadows. With reproductive technology there is no going back; its habit is propulsive, to advance ever further, evolving into a technique of ever greater instrumental value and refined efficacy.

A Bundle of Tissues

There is an inherent contradiction in human cloning: the very process is an exercise in dehumanization. By negating the inviolability of the human body, cloning is an intrusion into the *primum mobile* of the genetic ecosystem. The invasive vigor of this procedure will reach a truly awesome level when computers are placed in the service of cloning.

The moral and ethical impasse born out of cloning has many layers. Like the test tube baby that accomplishes human reproduction without sexual intercourse, cloning replaces procreation with replication and further confuses the functions of the human family.

Cloning raises all the age-old questions about life, but in a new mold. Is our body only a bundle of genes, tissues and organs? What is a person? What is the relation of the body to the spirit? In the Cartesian duality of mind and matter, how far can one go denying the link between organic composition and existential identity? Most worrisome of all, cloning forecloses genetic variability. It reinforces the values of genetic determinism because it poses a threat to individuality and diversity through identical replication.

Here the good old nature-nurture debate is in for a real shock! Cloning imposes a deterministic blueprint of bodily development, but cannot furnish the nurturing component of the person. In no small measure genetic determinism is the antithesis of moral and ethical choice. Where, with cloning, is the boundary between nature and nurture when the pre-selective hand of man reaches into the variety of combinant possibilities and makes his choice?

Choice brings us to the much-contested debate on parental vs. fetal rights. The issue gets even murkier than with the early stages of embryonic development when the long arm of *in utero* genetic manipulation becomes part of the picture. For instance, we can be nothing but mute on the risk of inherited disorders and the ability to fight disease in a person born of a frozen-and-thawed cloned embryo. Similarly, do parents have a right to deliberately alter the genetic endowment of a future child? Will that future child make a retroactive claim for damages inflicted through prebirth genetic selection?

And what about the market? In the biological bazaar, one is naggingly familiar with shopping for commodities like blood, sperm, ovum, or organs. Cloning gives a new meaning to the human body as merchandise. Instead of being content with the organ parts it would acquire novel techniques for wholesaling, packaging, and marketing made-to-order clones. In its instrumental mode, cloning will become an agent of commercial exploitation very much like the rent-a-womb syndrome from which we already suffer.

If success with the transgenic animals—where defective genes have been replaced so as to prevent the symptoms of an inherited disease—is any yardstick, then there is nothing whimsical about the idea of conducting business through a mail-order catalog where the genetic map of possibilities offers wide consumer choice. The only question is, cash or charge?

The Qur'an on Clones

In the Muslim consciousness the body is the medial, or middle ground, where the world of spirit and matter meet. It is the pivot around which one's world revolves. In Islam, there is neither an idea of "rights" over one's body nor an "ownership" of the body in the Western sense of the word. For a Muslim, the body is a trust from God. It is neither a solely owned property nor a disposable commodity: hence the interdiction against suicide. The temporary possession of the body does not imply its ownership by the possessor. The ritual prayer one recites at the death of a person comes as a vivid reminder: "He alone grants life and deals death; and unto Him you all must return"(*Qur'an 10:56*).

Notwithstanding some Muslims whose mislaid zeal appears to portray the Qur'an as a book of human embryology, there are verses aplenty that point to a *normative* guidance on human creation. Let us read a sample: "We have created [every one of] you out of dust, then out of a drop of sperm, then out of a germ cell, then out of an embryonic lump complete [in itself] and yet incomplete, so that We might make [your origin] clear unto you. And whatever We will [to be born] We cause to rest in the [mother's] womb for a term set [by Us]" (*Qur'an 22:5*). Another verse reads: "Was he not once a [mere] drop of a sperm that had been split, and thereafter became a germ-cell—whereupon He created and formed [it] in accordance with what [it] was meant to be, and fashioned out of the two sexes, the male and the female? (*Qur'an 75:37-8*).

The Quranic paradigm of human creation, it would appear, preempts any move toward cloning. From the moment of birth to the point of death, the entire cycle is a divine act. Humankind is simply an agent, a trustee of God. The body is a trust from God. In the absence of a Quranic axiom on body as property, genetic intervention would appear to be quite unethical.

On the utilitarian side of the corporeal possession, Muslims are exhorted—as a ritualistic obligation—to keep this trust in good shape. Given that cloning is an asexual experience (in the sense that it is performed within the legal marital bonds; no extramarital genetic boundaries are crossed and the genetic endowment is only from the spouses), its prohibition must be judged against Islamic ethical norms. For instance, unlike Catholic strictures, Islam sanctions therapeutic abortion in cases of a genuine clinical condition, that is, impending danger to a mother's life.

Would cloning offer an analogous condition? We can think of only one possible scenario: prenatal corrective genetic intervention, provided there exists a clinical justification. Our reasoning for this assertion takes root in the body-as-trust paradigm and the ensuing responsibility for its care as the duty of every Muslim woman and man.

The arrogance of Western science has never been greater than when it crossed the boundary of cloning. Does cloning represent the malevolence of the rebellious? Is it the vengeful self-perpetuation of those who would defy God? The human body is God's property, not man's laboratory. To abuse God's trust will only lead to a travesty of the human essence.

Buddhism and Cloning

Courtney Campbell

The Buddhist Churches in America claim approximately 100,000 adherents. There are, in addition, numerous non-affiliated Buddhist temples, monasteries, and organizations. There is as yet no systematic consideration of cloning by Buddhist scholars, nor is there any formal teaching authority. This manifests the Buddha's warning to his followers that speculation about metaphysical issues was futile because the human problems of birth, old age, death, and sorrow remain regardless. However, basic Buddhist teachings present an ethic of responsibility, centered on the values of non-injury and the relief of suffering of sentient beings, compassion, the "no-self," the moral authority of intuition, and reincarnation. These values offer some elements of a Buddhist response to reproductive and genetic technologies, including cloning.

Buddhist teachings indicate that the Buddha (fl. 5th c. BCE) provided a fourfold decision-making method for his followers should they encounter unanticipated questions. The four steps involve recourse to (1) original Buddhist texts; (2) derivation of rules in "consonance" with the original texts; (3) the views of respected teachers; (4) the exercise of personal judgement, discretion, and opinion. Buddhist scholars have cited this method as a resource for Buddhists in addressing the issues of cloning, with a particular emphasis on the authoritative nature of personal intuition and opinion (Nakasone). By its nature, then, there is a notable diversity of views by Buddhists on cloning, rather than a Buddhist view.

Procreation and Reproduction.

Buddhist scholars generally agree that the process by which children are born into the world makes no difference. "Individuals can begin their lives in many ways," including but not limited to human sexual generation. Cloning is thereby understood as an alternative method of generating new human life, in principle continuous with other methods (Keown). One Buddhist ethicist has supported use of reproductive technology, so long as it benefits the couple who wish to have a child and does not bring pain or suffering. However, some Buddhist scholars find in human cloning an impoverished approach to procreation. It marks a diminished creativity and diversity, analogous to the difference between the creativity, initiative, and investment that is required for an original painting and the mechanistic process required to reproduce the painting (Nolan).

Human Status and Enlightenment.

The status of human being is critical within Buddhist thought, because it is the only ontological condition by which an entity can achieve "enlightenment" and liberation from a world marked by suffering. Buddhist scholars throughout history have reiterated that, due to *karma*, the chances of being born as a human being are rare and remote. Human life is a precious opportunity to escape from perpetual rebirth (*karma-samsara*) by following the teachings (*dharma*) of the Buddha.

In this respect, any form of human reproduction, sexual or asexual, that allows for the birth of a human being may be especially valuable. Buddhist tradition contains stories of "spontaneous generation." Buddhist scholar Damien Keown states that cloning, if it "is ever perfected in human beings, would show only that there are a variety of ways in which life can be generated. It would not cast doubt on whether the host from which the clone was taken, or the clone itself, were ontological individuals" (Keown, 90).

Some forms of Buddhism may endorse cloning because of the chance human life gives to achieve enlightenment. The Dalai Lama, the exiled leader of Tibetan Buddhism, was questioned about his attitude towards the following hypothetical scenario: "[What] if at some future time...you could make by genetic engineering, with proteins and amino acids, or by engineering with chips and copper wires, an organism that had all of our good qualities and none of our bad ones...?" The Dalai Lama indicated he would welcome such a technological development, because it would facilitate the process of rebirth and liberation.

Moral Development and Spiritual Priorities.

Buddhist understandings that change is the nature of reality suggest that, in considering technological developments, the central questions concern how persons can accommodate change and how they can use change to expand their self-understanding and their understanding of humanity. Cloning may be an occasion for self-knowledge, which is a central feature to the experience of enlightenment. Nonetheless, the end of enlightenment as an end in itself may not, for some Buddhists, justify the use of any means of reproduction.

A different position on cloning can be supported by claims

and stories in Buddhist texts. It is important in Buddhism for children to express generosity to their parents, especially the mother, for the risks of birth and nurture they assume in bringing a child into the world. Human cloning offers a way of reproduction that, if efficient, would diminish risk, and thus diminish the generosity and gratitude of the child.

Moreover, while cloning may preserve genetic identity, it cannot assist in what for Buddhists is most critical—the cultivation of spiritual identity. The problem of distorted priorities is illustrated in a famous narrative, the "Parable of the Mustard Seed." In the parable, a distraught woman sought out the Buddha, requesting that he restore life to her dead child. The Buddha indicated that a cure was simple: The woman needed to prepare tea from five or six grains of mustard seed. The Buddha stipulated, however, that the grains needed to come from a house not visited by death. The woman was unable to obtain a single grain, thus learning about the universal truth of death. This narrative supports Buddhist concerns with cloning research or human cloning due to the attention focused on bodily, material life to the neglect of cultivating discovery or the inner life of a person. This misguided priority is reflected in the statement of Gen Kelsang Tubpa, a Buddhist monk: "Cloning is just another example of man's belief that by manipulating the external environment he will create happiness for himself and freedom from suffering."

Some Buddhist scholars have raised objections to applications of cloning, particularly commercial or social agendas that may support cloning for reasons contrary to the interest of the clone. These agendas may include pressures on scientists for continual progress and discovery or for commercial gain from pharmaceuticals or organ harvesting. In this respect, there would be greater suspicion within Buddhism about private-sponsored cloning research without public oversight.

Sentience and Cloning Research.

While cloning might be permissible under some understandings of Buddhism, the scientific research necessary to build up to cloning encounters difficulties. Part of the "Noble Eightfold Path" promulgated by the Buddha prohibits infliction of violence or harm on *sentient* beings. This would seem to permit research on human pre-embryos, but the primacy Buddhism places on birth as a human being as a necessary condition of enlightenment can restrict such research. Buddhism does hold that a new being comes into existence shortly after fertilization. Moreover, especially where the research process is very inefficient and causes loss of life, both embryo research and animal research would be especially problematic. Any Buddhist account would ask of cloning research or human cloning: "How does this serve all sentient beings?"

References

Damien Keown, *Buddhism and Bioethics* (New York: St. Martin's Press, 1995).

Nolan, K., "Buddhism, Zen, and Bioethics," in B.A. Lustig (ed.), *Theological Developments in Bioethics: 1990–1992* (Dordrecht: Kluwer Academic Publishers, 1993), pp. 185–216.

The Ethics and Politics of Small Sacrifices in Stem Cell Research

Glenn McGee and Arthur Caplan

Pluripotent Human Stem Cell research may offer new treatments for hundreds of diseases (Thomson et al. 1998; Vogel 1999), but opponents of such research argue that pluripotent stem cell therapy comes attached to a Faustian bargain: the destruction of many frozen embryos for every new cure. The National Bioethics Advisory Commission (NBAC), the Geron Ethics Advisory Board (GEAB), and many scholars of bioethics, including John Robertson, have raised interesting ethical issues about potential stem cell research (NBAC 1999; GEAB 1999; McGee and Caplan 1999; Fletcher 1999; Robertson 1999). But nothing holds the attention of the public or lawmakers like the brewing battle between millions of ill Americans who favor stem cell research and millions who oppose the destruction of any fetus or embryo for any purpose. The plan to sacrifice embryos for a revolutionary new kind of research has reawakened a long-dormant academic debate about the morality of destroying developing human life.

Robertson (like the GEAB and NBAC) argues that one's position on whether pluripotent stem cells can be used in research depends on one's view of the intrinsic and symbolic status of the embryo (Robertson 1999). In essence, Robertson and many others argue that one's position on stem cell research pretty much boils down to one's position on abortion. But that is too simple a framework.

Robertson and NBAC make two mistakes: first, they duck the task of identifying and analysing criteria for justifying the sacrifice of human cells and human life in the name of research. All who write about pluripotent stem cells refer to respect for pro-life views about symbolic moral status, and GEAB in particular claims that symbolic status is the reason such research should be therapeutic in nature. Yet so far no one has explained what kind of destruction is appropriate and of what sort of creatures and for what specific goals and contexts.

Second, Robertson, NBAC, and GEAB take what we call an "accommodationist" posture toward political opposition to stem cell research. In their haste to avoid rehashing the abortion debate, many proponents of stem cell research, including the National Institutes of Health Counsel, Harriet Raab (1999), cede too much ground to foes of abortion procedures, allowing too many arguments to go unanswered. They thus miss critical opportunities to engage in ethical debate about what sorts of rights and duties attend the making and management of embryonic life.

Moral Sacrifice

It seems to us that the central moral issues in stem cell research have less to do with abortion than with the criteria for moral sacrifices of human life. Those who inveigh against the derivation and use of pluripotent stem cells make the assumption that an embryo has not only the moral status of human person, but also a sort of super status that outweighs the needs of others in the human community. It is wrong to abort or kill a human, they argue, and thus it is wrong to kill an embryo. But this argument, which is problematic when made about abortion more generally, is doubly so when made against the derivation of pluripotent stem cells from embryos.

Even if frozen human embryos are persons, symbolically or intrinsically, this in no way entails the right of a frozen embryo to gestation, or to a risk-free pathway into maturation. Adult and child human beings' right "to life" is, considered constitutionally and as a moral problem, at best a negative right against unwarranted violence by the state or individuals. There are, sadly, few positive rights involved. Americans cannot, for example, claim a right against the state to protect them against disease, disasters, adverse weather, and other acts of nature. If a frozen human embryo is a full human person, it still has no right to life per se, but rather a negative right against unwarranted violence and a weak positive right to a set of basic social services (police protection, fire protection, and the like). The question remains as to what constitutes unwarranted violence against an embryo, and for what reasons might an embryo ethically be destroyed—e.g., in the interest of saving the community. Adults and even children are sometimes forced to give life, but only in the defense or at least interest of the community's highest ideals and most pressing interests. One would expect that the destruction of embryonic life, whatever its moral status, would also take place only under the most scrupulous conditions and for the best communal reasons. It bears noting that only those who consistently oppose all violence, destruction, or killing of any kind in the name of the state, the church, or the community can rationally oppose the destruction of an embryo solely by virtue of its status as a human person.

It remains to be shown what the common good is, and what sort of sacrifice an embryo should make in its interest. It is commonly held that no human being should be allowed to lie unaided in preventable pain and suffering.

The desire to ameliorate the suffering of the ill motivated Hippocrates, St. Francis of Assisi, Cicero, and Florence Nightingale. It is a central tenet of contemporary medicine that disease is almost always to be attended to and treated because it brings such pain and suffering to its victims and to their family and communities. Trade-offs are made in the treatment of disease, against cost and other competing social demands. But both the Western ethic of rescue and the practical structure of contemporary health care and other social institutions make it clear that among the deepest moral habits of human life is that of compassion for the sick and vulnerable. One of the compelling tenets of the movement to prevent abortion is the argument that a pregnancy ought not be terminated for superficial reasons, but should be viewed as a responsibility to aid the developing human life and to prevent it from needless suffering.

It is the moral imperative of compassion that compels stem cell research. Stem cell research consortium Patient's Cure estimates that as many as 128 million Americans suffer from diseases that might respond to pluripotent stem cell therapies. Even if that is an optimistic number, many clinical researchers and cell biologists hold that stem cell therapies will be critical in treating cancer, heart disease, and degenerative diseases of aging such as Parkinson's disease. More than half of the world's population will suffer at some point in life with one of these three conditions, and more humans die every year from cancer than were killed in both the Kosovo and Vietnam conflicts. Stem cell research is a pursuit of known and important moral goods.

What is Destroyed?

The sacrifice of frozen embryos is a curious matter. Set aside, again, the question of whether a frozen embryo is a human life or human person. Grant for a moment that a 100-cell human blastocyst, approximately the size of the tip of an eyelash and totally lacking in cellular differentiation, is a fully human person. What does such a person's identity mean, and in what ways can it be destroyed? What would it mean for such a person to die? When could such a death be justified? These questions require a new kind of analysis.

The human embryo from which stem cells are to be taken is an undifferentiated embryo. It contains mitochondria, cytoplasm, and the DNA of mother and father within an egg wall (which also contains some RNA). None of the identity of that embryo is wrapped up in its memory of its origins: it has no brain cells to think, no muscle cells to exercise, no habits. The 100-cell embryo has one interesting and redeeming feature, which as best anyone can tell is the only thing unique about it: its recombined DNA. The DNA of the embryo contains the instructions of germ cells from father and mother, and the earliest moments of its conception determined how the DNA of mother and father would be uniquely combined into a human person. The DNA of that person will, if the embryo survives implantation, gestation, and birth, continue to direct many facets of the growth and identity of our human person. At 100 cells, nuclear DNA is the only feature of the embryo that is not replaceable by donor components without compromising the critical features of the initial recombination of maternal and paternal genetic material after sex (or, in this case, *in vitro*).

Opponents of stem cell research make an anti-abortion argument, namely that the harvesting of pluripotent stem cells will require the destruction of the embryo. But while the cytoplasm, egg wall, and mitochondria of the embryo are destroyed, we just noted that none of these cellular components identifies the embryo at the 100-cell stage. The personifying feature of a 100-cell blastocyst is its DNA. Pluripotent stem cells from the harvested embryo are directed to form cell lines, each cell of which contains, in dormant form, the full component of embryonic DNA. The DNA in the cell lines has a much greater chance of continuing to exist through many years than does the DNA of a frozen embryo (which in most cases already will have been slated for destruction by the IVF clinic that facilitated the donation, and which would have no better than a 5 to 10 percent chance of successful implantation in any event). Although most Americans are opposed to the "cloning" of adult human beings, it might be possible to harvest DNA from any of the stem cell–based cell lines to make a new, nuclear transfer–derived embryo, or in fact to make five or ten embryos, each of which would possess all of the DNA of the original embryo. In this sense, the critical, identifying features of the embryo would never have been destroyed in the first place, unless what one means by "destroying" an embryo is the loss of its first egg wall, cytoplasm, and mitochondria. The transfer of the nucleus from an embryo to an enucleated egg is a bit like a transplant, though here the donor and the donation are both the DNA. In the case of embryos already slated to be discarded after IVF, the use of stem cells may actually lend permanence to the embryo. Our point here is that the sacrifice of an early embryo, whether it involves a human person or not, is not the same as the sacrifice of an adult, because the life of a 100-cell embryo is contained in its cells' nuclear material.

The task of balancing sacrifice in the community is one encountered by Solomon in the Judeo-Christian religious texts. Our institutions must enable us in the community to debate and identify the ideals that merit sacrifice, and the loss must be weighed with justice in mind. An embryo cannot reason and it cannot reject a sacrifice or get up and leave the community. For those who feel special responsibility to embryos, the vulnerability of the frozen embryo may suggest special consideration of the kind given to all moral actors in society who are for one reason or another without voice. The question remains, though: what need is so great that it rises to the level where every member of the human family might sacrifice for it? Already it is clear that we believe that no need is more obvious or compelling than the suffering of half the world at the hand of disease. Not even the most insidious dictator could dream up a chemical war campaign as horrific as the devastation wrought by Parkinson's disease, which destroys our grandparents, parents, and finally many of us.

If Parkinson's disease, or for that matter if a dictator could only be stopped through the destruction of an infant, most every human would blanch at the idea of such sacrifice. But that is not what is asked here. Even those who hold that an embryo is a person will not want to argue that the life of a 100-cell embryo is contained in its inessential components. Assuming that a developing embryo can be salvaged by transplanting its DNA, as we have described, it seems unreasonable to oppose the destruction of the embryo's external cellular material or to fear that the 100-cell embryo is killed in the transfer. The identity of the human embryo person, if it is a symbolic or intrinsic person, is tied at that stage to the DNA. That DNA is not lost or even injured by the harvesting of embryonic stem cells. This is not the sacrifice of the smallest and most vulnerable among us. We are debating the potential for temporary transplant of undifferentiated tissue and the DNA of such a "person," rather than the imminent destruction of an embryo discarded by a clinic. It is difficult to imagine those who favor just war opposing a war against such suffering given the meager loss of a few cellular components.

Political Accommodation

The road to a democratic debate about stem cell research is difficult, as Robertson acknowledges. We entirely agree with Robertson that NIH, and researchers working on stem cell–based cell lines, need not be considered complicit in the destruction of an embryo far removed in time and space (Robertson 1999). However, complicity assumes someone has done something wrong. Critics and NBAC are thus right to call attention to the amoral nature of the Raab opinion. It could not be more plain that NIH (and some politicians who favor stem cell research) initially hoped to sidestep entirely any ethical debate over the use of cell lines derived from the destruction of embryos. As we have argued above, it hardly does justice to those who believe that embryos are human persons to sacrifice those persons in a cloud of deception and cowardice. Moral sacrifice demands the highest accountability of our social institutions and especially those entrusted with writing and litigating the theory and policy that guides bodies like NIH.

Since at least 1980, the market on "family values" has been cornered by one side of the political spectrum. Anticipating that the public prefers platitudes about abortion to complex positions, most in American political office do not even discuss abortion and embryo research beyond a simple reference to their pro- or anti-abortion voting record. Meanwhile, the last twenty years have seen conservatives develop extraordinarily rigid and arcane positions on virtually every area of "family values."

Again and again, bioethicists' and politicians' policy of appeasing conservative views on family values has resulted in bad policy and balkanized public debate. Embryo and fetal tissue research are particularly instructive in this regard. Without an engagement of the moral questions involved, and without an explicit moral framework to explain the decisions, the nation found itself with bans on embryo research and fetal tissue transplantation research that the majority of citizens did not support.

NIH Counsel Raab, National Institutes of Health Director Harold Varmus, NBAC, the GEAB, and Robertson each give too much moral ground by ceding the illogical (embryos are special people who can never be allowed to die) and bizzare (researchers must not be complicit in the death of an embryo), arguments made by opponents of stem cell research. The sight of scientists rushing to find a legal opinion that legitimizes their work is not a pretty one, nor is that of politicians running scared from those who threaten to shut down the National Institutes of Health in the interest of preserving frozen embryos in stasis. We must be conscious of the role of politics (McGee 1999), but the law and the facts do not help with stem cell research. The issues here are novel and they are hard, but mostly they require philosophical innovation about what an embryo is and how we are to treat embryonic material in a time of stem cell research. Our argument here is that no embryo need be sacrificed, but we must alter the terms and goals of our debate to frame an appropriate moral framework for dealing with embryos.

Notes

[1] Fletcher, John. 1999. "Deliberating Incrementally on Human Pluripotential Stem Cell Research." National Bioethics Advisory Commission Background Papers on Embryonic Stem Cell Research.

[2] Geron Ethics Advisory Board. 1999. "Research with Human Embryonic Stem Cells: Ethical Considerations." *Hastings Center Report* 29 (2): 36–38.

[3] McGee, Glenn. 1999. Pragmatic Method in Bioethics. In *Pragmatic Bioethics*, ed. Glenn McGee, pp. 31–73 (Nashville: Vanderbilt University Press).

[4] *Idem*, and Arthur Caplan. 1999. "What's in the Dish?" [Symposium on Issues in Stem Cell Research] *Hastings Center Report* 29 (2): 38–41.

[5] NBAC. National Bioethics Advisory Commission. 1999. *The Ethical Use of Human Stem Cells in Research*. Draft Report (15 June). Rockville, MD: NBAC.

[6] Raab, Harriet. 1999. Memorandum to Harold Varmus, M.D., Director, NIH, Federal Funding for Research Involving Human Pluripotent Stem Cells, 15 January.

[7] Robertson, John A. 1999. "Ethics and Policy in Embryonic Stem

Cell Research." *Kennedy Institute of Ethics Journal* 9: 109–36.

[8] Thomson, James A.; Itskovitz-Eldor, Joseph; Shapiro, Sander; et al. 1998. "Embryonic Stem Cell Lines Derived from Human Blastocysts." *Science* 282: 1145–47.

[9] Vogel, Gretchen. 1999. "Harnessing the Power of Stem Cells." *Science* 283: 1432.

Please Don't Call It Cloning!

Bert Vogelstein, Bruce Alberts, Kenneth Shine

Scientists rely on a dialect of specialized terminology to communicate precise descriptions of scientific phenomena to each other. In general, that practice has served the community well—novel terms are created when needed to document new findings, behaviors, structures, or principles. The lexicon of science is constantly evolving. Scientists who are fluent in the language of any specific discipline can speak to one another using shorthand expressions from this dialect and can convey an exact understanding of their intended meanings, However, when the scientific shorthand makes its way to the nonscientific public, there is a potential for such meaning to be lost or misunderstood, and for the terminology to become associated with research or applications for which it is inappropriate.

In scientific parlance, cloning is a broadly used, shorthand term that refers to producing a copy of some biological entity—a gene, an organism, a cell—an objective that, in many cases, can be achieved by means other than the technique known as somatic cell nuclear transfer. Bacteria clone themselves by repeated fission. Plants reproduce clonally through asexual means and by vegetative regeneration.

Much confusion has arisen in the public. in that cloning seems to have become almost synonymous with somatic cell nuclear transfer, a procedure that can be used for many different purposes, Only one of these purposes involves an intention to create a clone of the organism (for example, a human). Legislation passed by the House of Representatives and under consideration in the U.S. Senate to ban the cloning of human beings actually proscribes somatic cell nuclear transfer—that is, any procedure in which a human somatic cell nucleus is transferred into an oocyte whose own nucleus has been removed. As Donald Kennedy remarked in a *Science* editorial last year, the legislation would interdict a wide range of experimental procedures that, in the near future, might become both medically useful and morally acceptable (1).

A law that would make it illegal to create embryonic stem cells by using somatic cell nuclear transfer would foreclose at least two important avenues of investigation. First, the technique shows promise to overcome the anticipated problem of immune rejection in stem cell-based therapies to replace a patient's diseased or damaged tissue. Creating stem cells with the patient's own nuclear genome might theoretically eliminate tissue rejection (2). Second, creating stem cell lines by using the somatic cell nuclei of individuals with heritable diseases offers an unprecedented opportunity to study genetic disorders as they unfold during cellular development.

Both of these research goals have nothing to do with producing a human being. They may be caught up in the proposed legislation in part because of misunderstood scientific jargon—namely, the casual use of the term "therapeutic cloning" to describe stem cells made for research in regenerative medicine using somatic cell nuclear transfer. What is worse, the already blurred distinction between these two very different avenues of investigation has been compounded by the interchangeable use of human cloning with therapeutic cloning by those who suggest that cloning a human being is a "therapeutic" treatment for infertility.

The term cloning, we believe, is properly associated with the ultimate outcome or objective of the research, not the mechanism or techniques used to achieve that objective. The goal of creating a nearly identical genetic copy of a human being is consistent with the term human reproductive cloning, but the goal of creating stem cells for regenerative medicine is not consistent with the term therapeutic cloning. The objective of the matter is not to create a copy of the potential tissue recipient, but rather to make tissue that is genetically compatible with that of the recipient. Although it may have been conceived as a simple term to help lay people distinguish two different applications of somatic cell nuclear transfer, "therapeutic cloning" is conceptually inaccurate and misleading, and should be abandoned.

	NUCLEAR TRANSLPLANTATION	HUMAN REPRODUCTIVE CLONING
END PRODUCT	Cells growing in a petri dish	Human being
PURPOSE	To treat a specific disease of tissue degeneration	Replace or duplicate a human
TIME FRAME	A few weeks (growth in culture)	9 months
SURROGATE MOTHER NEEDED	No	Yes
SENTIENT HUMAN CREATED	No	Yes
ETHICAL IMPLICATIONS	Similar to all embryonic cell research	Highly complex issues
MEDICAL IMPLICATIONS	Similar to any cell-based therapy	Safety and long-term efficacy concerns

Table: Crucial differences between cloning and transplantation

It is in the interest of the scientific community to clearly articulate the differences between stem cell research and human cloning. Most scientists agree that cloning a human being, aside from the moral or ethical issues, is unsafe under present conditions. A recently released National Academy of Sciences report details the considerable problems observed in the use of somatic cell nuclear transfer for animal reproduction and concludes that cloning of human beings should be prohibited (3). But the report also notes the substantial medical and scientific potential of stem cell lines created by using this technique.

More careful use of terminology would help the public and lawmakers sort out the substantial differences between nuclear transplantation and human reproductive cloning (table). One place to start is to find a more appropriate term for the use of somatic cell nuclear transfer to create stem cells. We propose the term "nuclear transplantation," which captures the concept of the cell nucleus and its genetic material being moved from one cell to another, as well as the nuance of "transplantation," an objective of regenerative medicine.

Legislators attempting to define good public policy regarding human cloning need the scientific community to be clear about the science, and to be clear when they speak to the public about it. Adopting the term nuclear transplantation in relation to stem cell research would be more precise, and it would help to untangle these two very different paths of investigation.

Notes

1 D. Kennedy, *Science* 294,745 (2001).

2 National Research Council Stem Cells and the Future of Regenerative Medicine (National Academy Press Washington, DC, 2001); available at wvw.nap.edu/

3 National Research Council, Scientific and Medical Aspects of Human Reproductive Cloning (National Academy Press, Washington, DC, 2002); available at www.nap.edu/

Remarks by the President on Human Cloning Legislation

George W. Bush

All of us here today believe in the promise of modern medicine. We're hopeful about where science may take us. And we're also here because we believe in the principles of ethical medicine. As we seek to improve human life, we must always preserve human dignity. And therefore, we must prevent human cloning by stopping it before it starts.

We live in a time of tremendous medical progress. A little more than a year ago, scientists first cracked the human genetic code—one of the most important advances in scientific history. Already, scientists are developing new diagnostic tools so that each of us can know our risk of disease and act to prevent them.

One day soon, precise therapies will be custom-made for our own genetic makeup. We're on the threshold of historic breakthroughs against AIDS and Alzheimer's Disease and cancer and diabetes and heart disease and Parkinson's Disease. And that's incredibly positive.

Our age may be known to history as the age of genetic medicine, a time when many of the most feared illnesses were overcome. Our age must also be defined by the care and restraint and responsibility with which we take up these new scientific powers.

Advances in biomedical technology must never come at the expense of human conscience. As we seek what is possible, we must always ask what is right, and we must not forget that even the most noble ends do not justify any means.

Science has set before us decisions of immense consequence. We can pursue medical research with a clear sense of moral purpose or we can travel without an ethical compass into a world we could live to regret. Science now presses forward the issue of human cloning. How we answer the question of human cloning will place us on one path or the other.

Human cloning is the laboratory production of individuals who are genetically identical to another human being. Cloning is achieved by putting the genetic material from a donor into a woman's egg, which has had its nucleus removed. As a result, the new or cloned embryo is an identical copy of only the donor. Human cloning has moved from science fiction into science.

One biotech company has already began producing embryonic human clones for research purposes. Chinese scientists have derived stem cells from cloned embryos created by combining human DNA and rabbit eggs. Others have announced plans to produce cloned children, despite the fact that laboratory cloning of animals has lead to spontaneous abortions and terrible, terrible abnormalities.

Human cloning is deeply troubling to me, and to most Americans. Life is a creation, not a commodity. Our children are gifts to be loved and protected, not products to be designed and manufactured. Allowing cloning would be taking a significant step toward a society in which human beings are grown for spare body parts, and children are engineered to custom specifications; and that's not acceptable. In the current debate over human cloning, two terms are being used: reproductive cloning and research cloning. Reproductive cloning involves creating a cloned embryo and implanting it into a woman with the goal of creating a child. Fortunately, nearly every American agrees that this practice should be banned. Research cloning, on the other hand, involves the creation of cloned human embryos which are then destroyed to derive stem cells.

I believe all human cloning is wrong, and both forms of cloning ought to be banned, for the following reasons. First, anything other than a total ban on human cloning would be unethical. Research cloning would contradict the most fundamental principle of medical ethics, that no human life should be exploited or extinguished for the benefit of another.

Yet a law permitting research cloning, while forbidding the birth of a cloned child, would require the destruction of nascent human life. Secondly, anything other than a total ban on human cloning would be virtually impossible to enforce. Cloned human embryos created for research would be widely available in laboratories and embryo farms. Once cloned embryos were available, implantation would take place. Even the tightest regulations and strict policing would not prevent or detect the birth of cloned babies.

Third, the benefits of research cloning are highly speculative. Advocates of research cloning argue that stem cells obtained from cloned embryos would be injected into a genetically identical individual without risk of tissue rejection. But there is evidence, based on animal studies, that cells derived from cloned embryos may indeed be rejected. Yet even if research cloning were medically effective, every person who wanted to benefit would need an embryonic clone of his or her own, to provide the designer tissues. This would create a massive national market for eggs and egg donors, and exploitation of women's bodies that we cannot and must not allow.

I stand firm in my opposition to human cloning. And at the same time, we will pursue other promising and ethical ways to relieve suffering through biotechnology. This year for the first time, federal dollars will go towards supporting human embryonic stem cell research consistent with the ethical guidelines I announced last August.

The National Institutes of Health is also funding a broad range of animal and human adult stem cell research. Adult stem cells which do not require the destruction of human embryos and which yield tissues which can be transplanted without rejection are more versatile than originally thought.

We're making progress. We're learning more about them. And therapies developed from adult stem cells are already helping suffering people.

I support increasing the research budget of the NIH, and I ask Congress to join me in that support. And at the same time, I strongly support a comprehensive law against all human cloning. And I endorse the bill—wholeheartedly endorse the bill—sponsored by Senator Brownback and Senator Mary Landrieu.

This carefully drafted bill would ban all human cloning in the United States, including the cloning of embryos for research. It is nearly identical to the bipartisan legislation that last year passed the House of Representatives by more than a 100-vote margin. It has wide support across the political spectrum, liberals and conservatives support it, religious people and nonreligious people support it. Those who are pro-choice and those who are pro-life support the bill.

This is a diverse coalition, united by a commitment to prevent the cloning and exploitation of human beings. It would be a mistake for the United States Senate to allow any kind of human cloning to come out of that chamber.

I'm an incurable optimist about the future of our country. I know we can achieve great things. We can make the world more peaceful, we can become a more compassionate nation. We can push the limits of medical science. I truly believe that we're going to bring hope and healing to countless lives across the country. And as we do, I will insist that we always maintain the highest of ethical standards.

April 20, 2002

Remarks by the President on Stem Cell Research

George W. Bush

Good evening. I appreciate you giving me a few minutes of your time tonight so I can discuss with you a complex and difficult issue, an issue that is one of the most profound of our time.

The issue of research involving stem cells derived from human embryos is increasingly the subject of a national debate and dinner-table discussions. The issue is confronted every day in laboratories as scientists ponder the ethical ramifications of their work. It is agonized over by parents and many couples as they try to have children, or to save children already born.

The issue is debated within the church, with people of different faiths, even many of the same faith coming to different conclusions. Many people are finding that the more they know about stem cell research, the less certain they are about the right ethical and moral conclusions.

My administration must decide whether to allow federal funds, your tax dollars, to be used for scientific research on stem cells derived from human embryos. A large number of these embryos already exist. They are the product of a process called in vitro fertilization, which helps so many couples conceive children. When doctors match sperm and egg to create life outside the womb, they usually produce more embryos than are planted in the mother. Once a couple successfully has children, or if they are unsuccessful, the additional embryos remain frozen in laboratories.

Some will not survive during long storage; others are destroyed. A number have been donated to science and used to create privately funded stem cell lines. And a few have been implanted in an adoptive mother and born, and are today healthy children.

Based on preliminary work that has been privately funded, scientists believe further research using stem cells offers great promise that could help improve the lives of those who suffer from many terrible diseases—from juvenile diabetes to Alzheimer's, from Parkinson's to spinal-cord injuries. And while scientists admit they are not yet certain, they believe stem cells derived from embryos have unique potential.

You should also know that stem cells can be derived from sources other than embryos—from adult cells, from umbilical cords that are discarded after babies are born, from human placenta. And many scientists feel research on these types of stem cells is also promising. Many patients suffering from a range of diseases are already being helped with treatments developed from adult stem cells.

However, most scientists, at least today, believe that research on embryonic stem cells offer the most promise because these cells have the potential to develop in all of the tissues in the body.

Scientists further believe that rapid progress in this research will come only with federal funds. Federal dollars help attract the best and brightest scientists. They ensure that new discoveries are widely shared at the largest number of research facilities and that the research is directed toward the greatest public good.

The United States has a long and proud record of leading the world toward advances in science and medicine that improve human life. And the United States has a long and proud record of upholding the highest standards of ethics as we expand the limits of science and knowledge. Research on embryonic stem cells raises profound ethical questions, because extracting the stem cell destroys the embryo, and thus destroys its potential for life. Like a snowflake, each of these embryos is unique, with the unique genetic potential of an individual human being.

As I thought through this issue, I kept returning to two fundamental questions: First, are these frozen embryos human life, and therefore, something precious to be protected? And second, if they're going to be destroyed anyway, shouldn't they be used for a greater good, for research that has the potential to save and improve other lives?

I've asked those questions and others of scientists, scholars, bioethicists, religious leaders, doctors, researchers, members of Congress, my Cabinet, and my friends. I have read heartfelt letters from many Americans. I have given this issue a great deal of thought, prayer and considerable reflection. And I have found widespread disagreement.

On the first issue, are these embryos human life—well, one researcher told me he believes this five-day-old cluster of cells is not an embryo, not yet an individual, but a pre-embryo. He argued that it has the potential for life, but it is not a life because it cannot develop on its own.

An ethicist dismissed that as a callous attempt at rationalization. Make no mistake, he told me, that cluster of cells is the same way you and I, and all the rest of us, started our lives. One goes with a heavy heart if we use these, he said, because we are dealing with the seeds of the next generation.

And to the other crucial question, if these are going to be destroyed anyway, why not use them for good purpose—I also found different answers. Many argue these embryos are byproducts of a process that helps create life, and we should allow couples to donate them to science so they can be used for good purpose instead of wasting their potential. Others will argue there's no such thing as excess life, and the fact that a living being is going to die does not justify experimenting on it or exploiting it as a natural resource.

At its core, this issue forces us to confront fundamental questions about the beginnings of life and the ends of science. It lies at a difficult moral intersection, juxtaposing the need to protect life in all its phases with the prospect of saving and improving life in all its stages.

As the discoveries of modern science create tremendous hope, they also lay vast ethical mine fields. As the genius of science extends the horizons of what we can do, we increasingly confront complex questions about what we should do. We have arrived at that brave new world that seemed so distant in 1932, when Aldous Huxley wrote about human beings created in test tubes in what he called a "hatchery."

In recent weeks, we learned that scientists have created human embryos in test tubes solely to experiment on them. This is deeply troubling, and a warning sign that should prompt all of us to think through these issues very carefully.

Embryonic stem cell research is at the leading edge of a series of moral hazards. The initial stem cell researcher was at first reluctant to begin his research, fearing it might be used for human cloning. Scientists have already cloned a sheep. Researchers are telling us the next step could be to clone human beings to create individual designer stem cells, essentially to grow another you, to be available in case you need another heart or lung or liver.

I strongly oppose human cloning, as do most Americans. We recoil at the idea of growing human beings for spare body parts, or creating life for our convenience. And while we must devote enormous energy to conquering disease, it is equally important that we pay attention to the moral concerns raised by the new frontier of human embryo stem cell research. Even the most noble ends do not justify any means.

My position on these issues is shaped by deeply held beliefs. I'm a strong supporter of science and technology, and believe they have the potential for incredible good—to improve lives, to save life, to conquer disease. Research offers hope that millions of our loved ones may be cured of a disease and rid of their suffering. I have friends whose children suffer from juvenile diabetes. Nancy Reagan has written me about President Reagan's struggle with Alzheimer's. My own family has confronted the tragedy of childhood leukemia. And, like all Americans, I have great hope for cures.

I also believe human life is a sacred gift from our Creator. I worry about a culture that devalues life, and believe as your President I have an important obligation to foster and encourage respect for life in America and throughout the world. And while we're all hopeful about the potential of this research, no one can be certain that the science will live up to the hope it has generated.

Eight years ago, scientists believed fetal tissue research offered great hope for cures and treatments—yet, the progress to date has not lived up to its initial expectations. Embryonic stem cell research offers both great promise and great peril. So I have decided we must proceed with great care.

As a result of private research, more than sixty genetically diverse stem cell lines already exist. They were created from embryos that have already been destroyed, and they have the ability to regenerate themselves indefinitely, creating ongoing opportunities for research. I have concluded that we should allow federal funds to be used for research on these existing stem cell lines, where the life and death decision has already been made.

Leading scientists tell me research on these sixty lines has great promise that could lead to breakthrough therapies and cures. This allows us to explore the promise and potential of stem cell research without crossing a fundamental moral line, by providing taxpayer funding that would sanction or encourage further destruction of human embryos that have at least the potential for life.

I also believe that great scientific progress can be made through aggressive federal funding of research on umbilical cord placenta, adult and animal stem cells which do not involve the same moral dilemma. This year, your government will spend $250 million on this important research.

I will also name a President's council to monitor stem cell research, to recommend appropriate guidelines and regulations, and to consider all of the medical and ethical ramifications of biomedical innovation. This council will consist of leading scientists, doctors, ethicists, lawyers, theologians and others, and will be chaired by Dr. Leon Kass, a leading biomedical ethicist from the University of Chicago.

This council will keep us apprised of new developments and give our nation a forum to continue to discuss and evaluate these important issues. As we go forward, I hope we will always be guided by both intellect and heart, by both our capabilities and our conscience.

August 9th, 2001

The Politics of Bioethics: Playing Defense Is Not Enough

Eric Cohen and William Kristol

"Nothing illustrates this administration's anti-science attitude better than George Bush's cynical decision to limit research on embryonic stem cells," declared John Kerry in a December 2003 campaign speech. He was referring to the president's August 9, 2001, decision to permit federal funding for existing embryonic stem cell lines, where the embryos in question had already been destroyed, but to deny funding for research involving further embryo destruction.

Ever since President Bush announced his stem cell policy, research advocates have attacked it as "not enough." They want more funding for more lines, without restrictions. They want the freedom to produce embryonic stem cell lines indefinitely, using as many embryos as necessary to advance research on a long list of terrible diseases. The idea of limits—in this case, no taxpayer funding for new embryo destruction—strikes them as incomprehensible and indefensible. In this spirit, Kerry attacks the Bush administration's "recessive gene of pessimism about progress and people," and declares that when "faced with a basic decision on America's health, George Bush chose to go to the right wing instead of the right way." Kerry aims to portray the Democrats as the party of health and progress, the Republicans as the party of suffering, death, and religious zeal.

The question is: How will President Bush respond? No doubt he will defend his policy on federal funding. And no doubt he will argue that the eligible stem cell lines are "enough" to get the medical benefits we seek, and that the issue is fundamentally about "respecting human life," not using it as a means to even the noblest ends. But it is not clear that simply playing defense on this and other bioethics issues will succeed. Indeed, over 200 congressmen sent a letter to the president last week demanding that the current restrictions on federal funding of embryonic stem cell research be lifted. Furthermore, it is increasingly clear that limits on federal funding alone do not guarantee our successfully navigating the "vast ethical mine fields" that President Bush warned of in his stem cell speech. This means reexamining what we have learned in the bioethics fight since it began in earnest in 2001, and sketching what a realistic *offense* might look like in the months and years ahead.

Since the president announced his stem cell policy in August 2001, the science of the brave new world has continued apace—not just the destruction of human embryos on a growing scale, but the manipulation of human reproduction in radical new ways. In its latest report, *Reproduction and Responsibility*, the President's Council on Bioethics finds that the practice of assisted reproduction technology (ART) is largely unregulated. New baby-making technologies are introduced willy-nilly into clinical practice, with little research regarding their effects on the children produced with their aid. Because so many embryos are implanted all at once, nearly half the children born using ART are twins or triplets with disproportionately and often dangerously low birth weights. Some ART clinics already advertise cosmetic baby-making services—such as preimplantation genetic screening to choose the sex of one's child—and these services only promise to increase as our genetic knowledge expands. And it is the ART clinics and their patients that produce thousands of "excess" embryos each year—embryos that are frozen indefinitely or destroyed for research.

Meanwhile, in February 2004, South Korean scientists announced the creation of the first cloned human embryos to the blastocyst stage—the stage when they could be implanted in a woman's uterus to initiate a pregnancy or destroyed in the laboratory to harvest stem cells. The report in *Science* magazine sounds hauntingly like the "decanting room" in Aldous Huxley's *Brave New World*—systematic, precise, unrepentant about its use of women as egg factories and human embryos as raw materials. The South Koreans harvested 242 eggs from 16 women, tested 14 different cloning "protocols," developed 30 human embryos to the 100-cell stage, and destroyed them all to get a single stem cell line.

Just a few months earlier, researchers working with animals showed that it is possible to produce both eggs and sperm from embryonic stem cells, including eggs from male embryos and sperm from female embryos. This means that it might be possible, someday soon, to produce a human child with two male parents or two female parents—and even a human child whose mother, father, or both is a dead embryo. Still other researchers fused together male and female embryos to produce a genderless human hybrid. Chinese researchers have already produced chimeric clones using rabbit eggs and human DNA. And what now seems prosaic—the destruction of IVF embryos for their stem cells—is a growing practice, with a number of states (New Jersey, California) contemplating new public funding initiatives, and a number of universities (Harvard, Stanford) actively creating new embryo research institutes.

While this research has proceeded, the political debate on bioethics has stalled. President Bush's August 2001 decision established an important moral principle, but also left an ambiguous legacy. The moral principle is that society as a whole, using taxpayer money, will not endorse the destruction of human embryos for any purpose; and it will not create public incentives for embryo destruction in the future. Zealous critics have denounced the policy as the 21st-century equivalent of silencing Galileo—attacking the president directly for imposing his personal religious views on science, and often ignoring the fact that Congress, not the president, made the law that prohibits federal funding of embryo research. More sober critics have argued that because more stem cell lines have been produced since the president's decision, these new lines should also be eligible for funding. The "life and death decision," they argue, has once again already been made. But moving the date of eligibility would undermine the moral logic of the Bush policy. It would send the message that the date will keep moving, and that embryo destruction today will be publicly funded tomorrow.

But the Bush decision, while principled, is also a partial decision: It offers no practical proposal for limiting embryo research in the private sector, though it probably discourages some scientists from engaging in research that cannot get NIH funding. It does not confront the question of what to do about excess embryo creation in in vitro fertilization (IVF) clinics during fertility treatment, or what to do about the roughly 400,000 embryos now frozen "in storage." (Only 3 percent of these frozen embryos, by the way, have been made available by their parents for research purposes.) Finally, the Bush decision gives the nation a stake in the success of embryonic stem cell research as a whole, and probably benefits (indirectly) those who destroy embryos with private funds by advancing the field.

In the end, neither side in the embryo debate is happy

with the current policy: Embryo research opponents lament the ongoing destruction of embryos in the private sector; embryo research advocates resent the limits on funding. But both sides also fear that things could get worse than they are now—that is, funding limits could loosen (the conservative worry) or legal restrictions could tighten (the liberal worry). The difference, however, is that research supporters are on the offensive—lobbying Congress and the president to make the funding policy more liberal, and aggressively seeking funding in individual states. Embryo research opponents, by contrast, are on the defensive: trying to preserve the Bush policy, with little hope or expectation of banning embryo research in the private sector.

In the one area where conservatives have tried to set broader limits on biotechnology—human cloning—the political fight remains stalled. Since 2001, the cloning debate has been a battle between two competing bills: the Brownback bill and the Hatch-Feinstein bill. The Brownback bill would ban all human cloning, including the creation and destruction of cloned embryos for research. The Hatch-Feinstein bill would endorse the creation and use of cloned embryos for research, then mandate the destruction of all cloned embryos to prevent the production of cloned children. The Brownback bill is the best way to stop the creation of cloned children, by stopping this act at the very first step. And it would set an important precedent that we should not "create human life solely for research and destruction." The Hatch-Feinstein bill, by contrast, makes the American public an accomplice in this troubling practice, and it creates for the first time a class of human life—cloned embryos—that must by law be destroyed.

The case for the Brownback bill is as clear today as it has been for the last three years. But while the Brownback bill has passed in the House of Representatives twice, passage in the Senate is blocked. In the meantime, there remain no ethical limits on biotechnology of any kind: no limits on radical new ways of making babies (cloning and beyond) and no limits on the creation and destruction of human embryos or later-stage fetuses for research, so long as it is done with private money. We are left fighting for limits that may never come, and playing defense for a policy that only deals with one small piece of the brave new world problem. Perhaps it is time to be both more realistic and more ambitious—more realistic about what is possible now, and more ambitious in seeking limits that go beyond the issue of cloning and beyond restrictions on federal funding for embryonic stem cell research.

FOR THOSE WHO WORRY about where reproductive biotechnology is taking us, there are three fundamental concerns: the destruction of innocent life, the degradation of the family, and the threat of eugenics. Each one requires some elaboration.

The first concern is that in the desire to save human life and promote scientific progress, we will become callous towards life, using the weakest among us as tools to keep the stronger alive. This concern overlaps—both politically and morally—with the abortion issue. Both involve questions about the violability or inviolability of nascent human life, and what we are willing to endure or forgo to respect it. But embryo research is at once more defensible and more corrupting than abortion. It is more defensible because the goal is a humanitarian one (to ease suffering and cure disease rather than end a pregnancy), and because the early-stage embryos in question are so existentially puzzling. They are microscopic, developing, genetically complete human beginnings—not just any beginnings, but the beginnings of a particular human life. But they are created outside their natural environment in the human womb, and often left frozen for years in the IVF clinics where they are made. These embryos may be "one of us," but they don't seem like one of us. The moral transgression of embryo destruction, though real, is not so obvious, while the sick child or Parkinson's patient is obviously suffering.

For the very same reason, embryo research is potentially more corrupting than abortion. It is a fruit we seek, not a transgression we tolerate. It is a premeditated project, not a decision made in crisis. Only the most extreme pro-choice advocates see abortion as a "good" and abortionists as heroes. But embryo-based medicine, if it were possible, would quickly become "standard practice" for the entire society, with leading researchers winning Nobel Prizes and parents who reject it for their children seen as legally negligent. Once cures exist, we might quickly forget that there is a moral problem here at all. Late-stage abortion requires a greater willingness of mother and doctor to look away from the facts of what they are doing, because of the obvious humanity of the developed fetus. But embryo research, so closely tied to the modern medical project that we all esteem, could become a celebrated American way of life in a way that abortion has not.

The second concern about biotechnology involves the degradation of the family, and the possibility that new ways of making babies will undermine the relationship between parents and children. So far, we see this problem most clearly in our fears about human cloning. To clone a child is to wreak havoc on the ties that bind the generations; it is to make our twin brothers into sons and twin sisters into daughters. It is to impose our perverse self-love on innocent children. But cloning is only one part of a larger project to transform human procreation and the human family. This larger project aims to use our biological cleverness to make us into post-biological beings—to create a world where male and female no longer matter, and where welcoming the newborn child as a mystery gives way to genetic screening, selection, and quality control.

Ironically, what made this project possible in the first place was acting technologically on the desire of infertile couples to have a child of their own, flesh of their flesh. To fulfill this biological desire, we invented a way to initiate human life in the laboratory—a way to bring human origins into full human view, and thus make them available for manipulation and control. The first IVF child was born in 1978. Since then, many infertile couples have had children of their own, with IVF to thank for this blessing. But as a result, we also opened the door to new ways of making babies that undermine the very biological ties that IVF aimed to serve. Only by bringing the embryo outside the human body is it now possible to give birth to another couple's child; to have a child where the identity of the father is "anonymous"; or to contemplate women giving birth to genetic copies of themselves or two men having a child that is the fruit of their mixed genes. While of course not all families reflect the biological ties between the generations, there is a difference between adopting a child in need and creating an orphan by design.

Looking back, the significance of IVF cannot be overstated: It is the source of the embryos that are now available for research; it is the technological solution for couples seeking a biological child; and it is the crucial first step in transforming human procreation in radical new ways. Looking ahead, however, it is also clear that we stand at yet another major threshold. IVF, in most cases, still mimics nature—producing a child that is the fruit of a coupled male and female. The new ways of making babies, by contrast, radically depart from nature's sexual pairing, and they violate the family structure that has long imitated and civilized our given nature in the rearing of children.

The final concern about biotechnology is that our growing technical control over reproduction will open the door to a new eugenics—where parents pick and choose the genetic characteristics of their offspring, and society pressures families not to have genetically unfit children. The longtime fear of genetic engineering—superbabies made to order—is far-fetched. The real danger is something more subtle. It is using genetic information to choose babies with a greater *probability* of their being superior in ways we desire—that is, a greater probability of being tall, or athletic, or musical, or smart. It is not so much the tyrannical parent as the tentative-obsessive parent that is the problem—the parent who is unwilling to accept the child as given, but obsessed with trying to get the best child possible.

The problem with assisted reproduction today is that infertile couples sometimes put their future child in danger. The problem tomorrow will be that fertile parents, so hungry to have the child they want, will forgo natural reproduction for the clinic—where embryos can be created, screened, and tested in advance. Today, we abort children with genetic diseases. Tomorrow, we will select children with genetic advantages—with all the expectations and deformations that this new imposition of parental will introduces into child rearing.

At present, all of these practices remain unregulated and unrestricted in America: The use of genetic screening techniques to try to pick children with "superior" genotypes is ungoverned and unmonitored. Embryo destruction remains fully legal in the private sector, and a recent law passed in New Jersey protects the right of researchers to harvest later-stage fetuses as research tools. Revolutionary new ways of making babies are unhindered, including the now imminent possibility of using the South Korean "cookbook" (as one researcher called it) to try to clone a human child.

In thinking about how to govern this free for all, we have the benefit of the recent unanimous recommendations from the President's Council on Bioethics. The council calls for a ban on implanting human embryos into an animal uterus; a ban on producing embryos with human sperm and animal eggs or animal sperm and human eggs; a ban on initiating a pregnancy for research purposes; a ban on buying, selling, or patenting human embryos; a ban on destroying or harming embryos for research once they reach the 10–14 day stage of development; and a ban on radical new ways of producing a child, including "blastomere fusion" (which would create a child with four genetic parents, not two), conceiving a child whose father or mother is a dead embryo or aborted fetus, and human cloning.

It should be obvious that enacting such recommendations would be a great improvement over the laissez-faire status quo. But the recommendations involving embryo destruction and human cloning have been criticized by some pro-lifers on a number of grounds: for not going far enough, for accepting practices that are unacceptable, and for undermining the ethical clarity required for opposing the misdeeds of the biotechnology project. These criticisms are serious but not decisive. They force the question with which we began: What is a realistic conservative offense on bioethics issues? How does the president balance the steady support of the pro-life community—often the only reliable critics of the new practices—with the need to reach beyond the pro-life community to pass bioethics legislation? Is there wisdom in the partial limits proposed by the council? We believe there is, and that it becomes clear by taking up the two major pro-life criticisms directly.

The first criticism is that the council's recommendations separate reproductive cloning and research cloning, and propose a ban on reproductive cloning only. In doing so, the critics say, the council tacitly endorses the creation of cloned embryos for research; it offers another version of the Hatch-Feinstein bill that pro-lifers have been fighting against for three years. But this is incorrect.

The council's recommendations offer a way of banning reproductive cloning that differs from the two bills that have so far gone nowhere in Congress. When it comes to the dignity of the family, the council is more ambitious than the Brownback bill—banning not only cloning but a number of radical ways of making babies. But it does this by recommending a ban on the *creation* of cloned embryos (or other wrongfully produced embryos) *with the intent* of implanting them to begin a pregnancy. Such a law would not (like Hatch-Feinstein) mandate the destruction of any embryos. It would not (like Hatch-Feinstein) endorse the use of embryos for research, but rather preserve the status quo of public silence. The illegal act (unlike Hatch-Feinstein) would be embryo creation, if not all embryo creation. And it would allow the fight for the Brownback bill to continue in parallel, while banning a range of reproductive practices that everyone abhors.

The pro-life rejoinder is that silence means an implicit endorsement of cloned embryo research. And yet, as Leon Kass has pointed out, the Brownback bill, which aims to ban research on cloned embryos, is silent on the creation and destruction of IVF embryos for research. Of course, pro-lifers also reject this practice. They don't endorse it simply by not trying to ban it, and they don't imply that cloned embryos have a more sacred status than IVF embryos. Rather, they take aim at the evils they can limit in the real world, while remaining legislatively silent about the evils they cannot now stop. This is exactly what the council's recommendations do as well—protecting the dignity of human procreation, while remaining silent on the destruction of early embryos.

The second pro-life rejoinder is that by offering an alternative to the Brownback bill, the council recommendations will undermine ongoing efforts to pass the Brownback bill (or legislation like it in the states). They point to the fight in Nebraska, where a pro-embryo-research legislator introduced the council's recommendations verbatim in an effort to stop passage of the Brownback-style bill. But the fact that a pro-embryo research senator is willing to propose recommendations endorsed by pro-life council members like Robert George and Mary Ann Glendon suggests not a weakening of the pro-life side, but a possible movement of the pro-research side in a more conservative direction. Indeed, the Brownback strategy, by itself, may make pro-lifers *less* ambitious than they could be in conservative states, where they might ban all creation of human embryos for research, not just the creation of cloned embryos.

Another pro-life criticism of the council's recommendations is that banning the destruction of embryos for research once they reach the 10- to 14-day stage of development would implicitly endorse research on the earliest human embryos; it would suggest that the moral standing of developing human life changes at the 10- to 14-day line. But this argument seems to us to miss the wisdom of seeking partial—and principled—limits. To ban all embryo destruction after 10 to 14 days is the embryo research equivalent of a partial-birth abortion ban. The only difference is that instead of the 8- to 9-month fetus being given protected status under the law, it is the 10- to 14-day-old embryo. Imagine if a pro-abortion activist like Kate Michelman endorsed the proposition that all abortions after 10 to 14 days should be outlawed. Pro-lifers would be ecstatic. To enact a 10- to 14-day limit on embryo research would put in place the strongest legal protections of developing human life in the post-*Roe v. Wade* era. It would force the other side to accept that at least some embryos are morally and legally inviolable. And if those embryos are to be protected, why not others? It would shift the terms of the debate in a pro-life direction, and limit coming evils (like fetus farming) without betraying pro-life principles.

Certainly a total ban on cloning—indeed a total ban on embryo research—would be ideal from a pro-life perspective. But such bans do not seem forthcoming at the federal level. The status quo prevails—which is ultimately a victory for biotechnology without limits. What conservatives need, instead, is a realistic offense, and the council's recommendations are a good example of this approach, though one could imagine other initiatives along these lines as well.

The council offers limits that are much better than nothing—by preventing the destruction of some innocent human life, stopping new ways of degrading human procreation and family ties, and shutting down some gateways to a new eugenics.

We stand at a crucial moment in the debate about reproductive biotechnology—a moment like the late 1960s and early 1970s on abortion, or the early 1970s on in vitro fertilization. Despite the many ethical and legal precedents cutting in the opposite direction—towards a culture of autonomy without limits—there is a widespread consensus today against the most radical new ways of making babies and against harvesting fetuses for research. Reproductive freedom does not yet mean the right to have a child by any means possible. And even the most ardent supporters of embryo research still say they would never harm an embryo after 14 days of development. This broad consensus leaves open a door for enacting limits on the most dehumanizing uses of biotechnology, but it is a door that will not remain open forever.

The council's report lays the groundwork for setting such limits. It establishes the principle that not all science is good for the country, and that scientists, too, must answer to the deliberative judgment of the American people. If we act today to prevent some of the worst abuses of biotechnology, we will at least have begun to face the task before us, governing scientific progress in a democratic and moral way.

Contributors

BRUCE ALBERTS is president, National Academy of Sciences, and chair, National Research Council, Washington DC.

MUNAWAR AHMAD ANEES is Editor in Chief of *Periodica Islamica.*

RONALD BAILEY is science correspondent for *Reason* magazine and the editor of *Earth Report 2000: Revisiting the True State of the Planet.*

GEORGE W. BUSH is the 43rd President of the United States.

COURTNEY S. CAMPBELL is Professor in the Department of Philosophy, and Director of the Program for Ethics, Science, and the Environment, at Oregon State University.

ARTHUR CAPLAN has authored *Am I My Brother's Keeper, Due Consideration,* and other books. He is in the Department of Medical Ethics and Director of the Center for Bioethics at the University of Pennsylvania.

ERIC COHEN is Editor of the journal *The New Atlantis,* and Director of the Project on Biotechnology and American Democracy at the Ethics and Public Policy Center in Washington DC.

RONALD COLE-TURNER is the H. Parker Sharp Professor of Theology and Ethics at Pittsburgh Theological Seminary, and the editor of *Human Cloning: Religious Responses.*

JONATHAN R. COHEN is Associate Professor in the University of Florida Levin College of Law.

RICHARD DOERFLINGER is Deputy Director of the Secretariat for Pro-Life Activities, Unites States Conference of Catholic Bishops.

LEON EISENBERG is Professor Emeritus of Social Medicine at Harvard Medical School and editor most recently of *Implications of Genetics for Health Professional Education.*

AUTUMN FIESTER is Senior Fellow at the University of Pennsylvania's Center for Bioethics, and directs the Master of Bioethics Program in Penn's Department of Medical Ethics.

LEON KASS is the Addie Clark Harding Professor in the College and The Committee on Social Thought at the University of Chicago, and author, most recently, of *The Beginning of Wisdom: Reading Genesis*. He is Chairman of The President's Council on Bioethics.

CHARLES KRAUTHAMMER has won the Pulitzer Prize for his widely syndicated column, and appears regularly on Inside Washington and Fox News.

WILLIAM KRISTOL recently co-authored *The War Over Iraq: America's Mission and Saddam's Tyranny*. He is Editor of the journal *The Weekly Standard.*

VANESSA KUHN is a doctoral student in the Department of Health Policy and Management at the Johns Hopkins University.

GLENN MCGEE is Senior Fellow in the Center for Bioethics at the University of Pennsylvania, the author, most recently, of *Beyond Genetics: Putting the Power of DNA to Work in Your Life*, and Editor of *The American Journal of Bioethics*.

CHRIS MOONEY writes a monthly online column for *Skeptical Inquirer* magazine, and is a senior correspondent for *The American Prospect*.

SHERWIN NULAND is a clinical professor of surgery at Yale University, and has authored the National Book Award–winning *How We Die: Reflections on Life's Final Chapter*.

PASQUALE PATRIZIO is in the Department of Obstetrics and Gynecology and Male Infertility Program at the University of Pennsylvania Medical Center.

JOHN ROBERTSON has written widely on law and bioethics issues, including the book *Children of Choice: Freedom and the New Reproductive Technologies*. He holds the Vinson & Elkins Chair in Law at the University of Texas, Austin.

CLAIRE ROBERTSON-KRAFT is a recent graduate of the University of Pennsylvania.

KENNETH SHINE is President, Institute of Medicine, Washington DC.

GREGORY STOCK is author of *Redesigning Humans: Choosing our Genes, Changing our Future*, and director of the Program on Medicine, Technology, and Society at UCLA's School of Public Health in Los Angeles.

BERT VOGELSTEIN is at the Howard Hughes Medical Institute and the Kimmel Cancer Center at Johns Hopkins University, and chaired the recent National Research Council report on stem cells.

POTTER WICKWARE is a researcher in the Wang Lab, University of California at San Francisco, managing editor of the journal *Eukaryotic Cell*, and a frequent contributor on biotechnology to *Nature*, *The New York Times*, and other publications.

IAN WILMUT is a researcher at the Roslin Institute in Scotland and co-author with A. E. Schnieke, J. McWhir, A. J. Kind, and K. H. S. Campbell of "Viable offspring derived from fetal and adult mammalian cells", *Nature* vol. 385 (6619), February 27, 1997, pp. 810–13.

Permissions

"Cloning as a Reproductive Right" excerpted from "Liberty, Identity, and Human Cloning", published originally in 76 *Texas Law Review* 1371 (1998). Copyright 1998 by the Texas Law Review Association. Reprinted by permission.

"The Wisdom of Repugnance: Why We Should Ban the Cloning of Humans" reprinted by permission of *The New Republic*. © 1997, The New Republic, Inc.

"The Uncertain Art: Narcissus Looks Into the Laboratory" reprinted from *The American Scholar* Autumn 1998. Copyright © by Sherwin B. Nuland.

"Would Cloned Humans Really Be Like Sheep" reprinted from *The New England Journal of Medicine* 340.6 (2/11/99), 471–475. Copyright © 1999 Massachusetts Medical Society. All rights reserved.

"In God's Garden: Creation and Cloning in Jewish Thought" reprinted courtesy of Jonathan R. Cohen and *The Hastings Center Report*.

"Please Don't Call It Cloning!" reprinted with permission from *Science* vol. 295 (2/5/2002) p.1237. Copyright 2002 The American Association for the Advancement of Science.

"Crossing Lines" reprinted by permission of *The New Republic*. © 2002, The New Republic, Inc.

"Is Biomedical Research Too Dangerous to Pursue?" reprinted with permission from *Science* vol. 303 (2/20/2004). Copyright 2004 The American Association for the Advancement of Science.

"The Future Is Later" reprinted with permission from *The American Prospect* vol. 13, no.13 (7/13/2002). Copyright 2002 The American Prospect, Inc.

"The Ethics and Politics of Small Sacrifices in Stem Cell Research" reprinted with permission from *The Kennedy Institute of Ethics* vol. 9, issue 2 (June 1999), pp. 151–8. Copyright 1999 The Johns Hopkins University Press.

"The Politics of Bioethics" reprinted with permission from *The Weekly Standard* vol. 9, issue 33 (5/10/04). Copyright 2004 *The Weekly Standard*.